Comprehensive Asymmetric Catalysis
Supplement 2

Springer

Berlin
Heidelberg
New York
Hong Kong
London
Paris
Tokyo

Eric N. Jacobsen · Andreas Pfaltz · Hisashi Yamamoto (Eds.)

Comprehensive Asymmetric Catalysis

Supplement 2

With contributions by numerous experts

Springer

ERIC N. JACOBSEN
Department of Chemistry and Chemical Biology
Harvard University
12 Oxford Street
MA 02138 Cambridge, USA
e-mail: jacobsen@chemistry.harvard.edu

ANDREAS PFALTZ
Department of Chemistry
University of Basel
St. Johanns-Ring 19
CH-4056 Basel, Switzerland
e-mail: andreas.pfaltz@unibas.ch

HISASHI YAMAMOTO
Department of Chemistry
University of Chicago
5735 South Ellis Avenue
Chicago, IL 60637, USA
e-mail: yamamoto@uchicago.edu

ISBN 3-540-20983-2 Springer-Verlag Berlin Heidelberg New York

Cataloging-in-Publication Data applied for
Bibliographic information published by Die Deutsche Bibliothek
Die Deutsche Bibliothek lists this publication in the Deutsche Nationalbibliografie;
detailed bibliographic data is available in the Internet at <http:/dnb.ddb.de>.

Springer-Verlag is a part of Springer Science + Business Media

springeronline.com

© Springer-Verlag Berlin Heidelberg 2004
Printed in Germany

Typesetting: Data conversion by medio Technologies AG, Berlin
Cover: E. Kirchner, Heidelberg

Printed on acid-free paper   62/3020xv   5 4 3 2 1 0

Authors

Annette Bayer
Department of Chemistry
University of Tromsø
9037 Tromsø
NORWAY
e-mail: annette.bayer@chem.uit.no

Mark Lautens
Department of Chemistry
University of Toronto
Toronto, Ontario M5S 3H6
Canada
e-mail: mlautens@chem.utoronto.ca

Takeshi Ohkuma
Department of Chemistry and Research
Center for Materials Science
Nagoya University
Chikusa, Nagoya 464-8602
JAPAN

Kiyoshi Tomioka
Graduate School of Pharmaceutical Sciences
Kyoto University
Yoshida
Sakyo-ku
Kyoto 606– 8501
JAPAN
e-mail: tomioka@pharm.kyoto-u.ac.jp

Ronald L. Haltermann
Department of Chemistry and Biochemistry
620 Parrington Oval
University of Oklahoma
Norman OK 73019
USA
e-mail: rhalterman@ou.edu

Ryoji Noyori
Department of Chemistry and Research
Center for Materials Science
Nagoya University
Chikusa, Nagoya 464-8602
JAPAN
e-mail: noyori@chem3.chem.nagoya-u.ac.jp

Jean-François Paquin
Department of Chemistry
University of Toronto
Toronto, Ontario M5S 3H6
Canada
e-mail: jfpaquin@chem.utoronto.ca

Akira Yanagisawa
Department of Chemistry
Faculty of Science
Chiba University
Inage
Chiba 263– 8522
JAPAN
e-mail: ayanagi@scichem.s.chiba-u.ac.jp

Preface to the 2nd Supplement

We are pleased to make available the second supplement to our three volume reference work *Comprehensive Asymmetric Catalysis*.

Seven chapters which are already in the major reference work have been supplemented. A new chapter on Aminohydroxylation of Carbon-Carbon Double Bonds has been included.

Eric N. Jacobsen, Cambridge   January 2004
Andreas Pfaltz, Basel
Hisashi Yamamoto, Chicago

Preface to the 1st Supplement

We have been gratified to see that the *Comprehensive Asymmetric Catalysis* three volume set has been received with enthusiasm by the chemical community since its publication in 1999. As was easily anticipated, advances in asymmetric catalysis have continued at an explosive pace since then. Recognition of the impact of this field on chemistry has been evidenced both in practical terms by the application of asymmetric catalytic methods in a variety of new laboratory and industrial contexts, and quite visibly through the lofty recognition of to three of the pioneers of the field in the 2001 Nobel Prize.

In order to keep this reference work as useful and fresh as possible, our plan from the start was to provide supplementary volumes on a periodic basis. These would contain updates to chapters on topics where there is substantial recent progress, and new chapters on emerging topics. We are delighted to provide you here with the first of these supplements.

Eric N. Jacobsen, Cambridge   August 2003
Andreas Pfaltz, Basel
Hisashi Yamamoto, Chicago

Contents (Supplement 2)

The chapters whose numbers have a gray background are completely new.

Supplement to 5.2

Hydrogenation of Non-Functionalized
Carbon–Carbon Double Bonds
Ronald L. Haltermann 1

Supplement to 6.4

Hydroboration of Carbonyl Groups
Takeshi Ohkuma, Ryoji Noyori 7

Supplement to 20.1

Dihydroxylation of Carbon–Carbon Double Bonds
Annette Bayer 21

Chapter 20.2

Aminohydroxylation of Carbon–Carbon Double Bonds
Annette Bayer 43

Supplement to 24

Allylic Substitution Reactions
Jean-François Paquin, Mark Lautens 73

Supplement to 27

Allylation of C=O
Akira Yanagisawa 97

Supplement to 31.1

Conjugate Addition of Organometallics to Activated Olefins
Kiyoshi Tomioka 109

Supplement to 34.2

Protonation of Enolates
Akira Yanagisawa 125

Subject Index 133

Contents (Supplement 1)

The chapters whose numbers have a gray background are completely new.

Supplement to Chapter 6.1

Hydrogenation of Carbonyl Groups
Takeshi Ohkuma, Ryoji Noyori   1

Supplement to Chapter 6.2

Hydrogenation of Imino Groups
Takeshi Ohkuma, Ryoji Noyori   43

Supplement to Chapter 6.3

Hydrosilylation of Carbonyl and Imino Groups
Takeshi Ohkuma, Ryoji Noyori   55

Supplement to Chapter 14

Heck Reaction
Masakatsu Shibasaki, Erasmus M. Vogl, Takashi Ohshima.   73

Chapter 16.4

C–H Insertion Reactions, Cycloadditions and Ylide Formation
of Diazo Compounds
Huw M. L. Davies   83

Supplement to Chapter 26.1

Alkylation of C=O
Kenso Soai, Takanori Shibata   95

Supplement to Chapter 26.2

Alkylation of C=N
Keiko Hatanaka, Hisashi Yamamoto   107

Supplement to Chapter 28

Cyanation of Carbonyl and Imino Groups
Petr Vachal, Eric N. Jacobsen   117

Supplement to Chapter 29.3

Nitroaldol Reaction
Masakatsu Shibasaki, Harald Gröger, Motomu Kanai. 131

Chapter 29.4

Direct Catalytic Asymmetric Aldol Reaction
Masakatsu Shibasaki, Naoki Yoshikawa, Shigeki Matsunaga 135

Chapter 29.5

Mannich Reaction
Shū Kobayashi, Masaharu Ueno 143

Supplement to Chapter 31.2

Catalytic Conjugate Addition of Stabilized Carbanions
Masahiko Yamaguchi 151

Supplement to Chapter 34.1

Alkylation of Enolates
David L. Hughes 161

Supplement to Chapter 39

Combinatorial Approaches
Amir H. Hoveyda, Kerry E. Murphy 171

Chapter 43

Acylation Reactions
Elizabeth R. Jarvo, Scott J. Miller 189

Chapter 44

Metathesis Reactions
Amir H. Hoveyda, Richard R. Schrock 207

Subject Index 235

Contents (Vol. I–III)

Volume I

1   **Introduction**
    Andreas Pfaltz.   3

2   **Historical Perspective**
    Henri B. Kagan   9

3   **Basic Principles of Asymmetric Synthesis**
    Johann Mulzer   33

4   **General Aspects of Asymmetric Catalysis**

4.1 Non-Linear Effects and Autocatalysis
    Henri B. Kagan, T. O. Luukas.   101

5   **Hydrogenation of Carbon-Carbon Double Bonds**

5.1 Hydrogenation of Functionalized Carbon-Carbon Double Bonds
    John M. Brown   121

5.2 Hydrogenation of Non-Functionalized Carbon-Carbon Double Bonds
    Ronald L. Halterman   183

6   **Reduction of Carbonyl and Imino Groups**

6.1 Hydrogenation of Carbonyl Groups
    T. Ohkuma · R. Noyori   199

6.2 Hydrogenation of Imino Groups
    Hans-Ulrich Blaser, Felix Spindler   247

6.3 Hydrosilylation of Carbonyl and Imino Groups
    Hisao Nishiyama   267

6.4 Hydroboration of Carbonyl Groups
    Shinichi Itsuno   289

7   **Hydrosilylation of Carbon-Carbon Double Bonds**
    Tamio Hayashi   319

8   **Hydroalumination of Carbon-Carbon Double Bonds**
    Mark Lautens, Tomislav Rovis.   337

9   **Hydroboration of Carbon-Carbon Double Bonds**
    Tamio Hayashi   351

10   Hydrocyanation of Carbon-Carbon Double Bonds
     T.V. RajanBabu, Albert L. Casalnuovo. 367

11   Hydrocarbonylation of Carbon-Carbon Double Bonds
     Kyoko Nozaki 381

12   Hydrovinylation of Carbon-Carbon Double Bonds
     T.V. RajanBabu. 417

13   Carbometalation of Carbon-Carbon Double Bonds
     Amir H. Hoveyda, Nicola M. Heron 431

14   Heck Reaction
     Masakatsu Shibasaki, Erasmus M. Vogl. 457

Volume II

15   Pauson-Khand Type Reactions
     Stephen L. Buchwald, Frederick A. Hicks. 491

16   Cyclopropanation and C-H Insertion Reactions

16.1 Cyclopropanation and C-H Insertion with Cu
     Andreas Pfaltz. 513

16.2 Cyclopropanation and C-H Insertion with Rh
     Kevin M. Lydon, M. Anthony McKervey 539

16.3 Cyclopropanation and C-H Insertion with Metals
     Other Than Cu and Rh
     André B. Charette, Hélène Lebel. 581

17   Aziridination
     Eric N. Jacobsen 607

18   Epoxidation

18.1 Epoxidation of Allylic Alcohols
     Tsutomu Katsuki 621

18.2 Epoxidation of Alkenes Other than Allylic Alcohols
     Eric N. Jacobsen, Michael H. Wu. 649

18.3 Epoxide Formation of Enones and Aldehydes
     Varinder K. Aggarwal. 679

19   Oxidation of Sulfides
     Carsten Bolm, Kilian Muñiz, Jens P. Hildebrand. 697

20   Dihydroxylation of Carbon-Carbon Double Bonds
     Istvan E. Markó, John S. Svendsen. 713

21   C-H Oxidation
     Tsutomu Katsuki 791

22   Baeyer-Villiger Reaction
     Carsten Bolm, Oliver Beckmann 803

23   **Isomerization of Carbon-Carbon Double Bonds**
      Susumu Akutagawa.   813

24   **Allylic Substitution Reactions**
      Andreas Pfaltz, Mark Lautens   833

25   **Cross-Coupling Reactions**
      Tamio Hayashi   887

26   **Alkylation of Carbonyl and Imino Groups**

26.1 Alkylation of Carbonyl Groups
      Kenso Soai, Takanori Shibata   911

26.2 Alkylation of Imino Groups
      Scott E. Denmark, Olivier J.-C. Nicaise   923

27   **Allylation of Carbonyl Groups**
      Akira Yanagisawa.   965

28   **Cyanation of Carbonyl and Imino Groups**
      Atsunori Mori, Shohei Inoue   983

Volume III

29   **Aldol Reactions**

29.1 Mukaiyama Aldol Reaction
      Erick M. Carreira   997

29.2 Addition of Isocyanocarboxylates to Aldehydes
      Ryoichi Kuwano, Yoshihiko Ito   1067

29.3 Nitroaldol Reaction
      Masakatsu Shibasaki, Harald Gröger.   1075

30   **Addition of Acyl Carbanion Equivalents to
      Carbonyl Groups and Enones**
      Dieter Enders, Klaus Breuer.   1093

31   **Conjugate Addition Reactions**

31.1 Conjugate Addition of Organometallic Reagents
      Kiyoshi Tomioka, Yasuo Nagaoka.   1105

31.2 Conjugate Addition of Stabilized Carbanions
      Masahiko Yamaguchi.   1121

32   **Ene-Type Reactions**
      Koichi Mikami, Masahiro Terada   1143

33   **Cycloaddition Reactions**

33.1 Diels-Alder Reactions
      David A. Evans, Jeffrey S. Johnson   1177

33.2 Hetero-Diels-Alder and Related Reactions
      Takashi Ooi, Keiji Maruoka   1237

33.3 [2+2] Cycloaddition Reactions
      Yujiro Hayashi, Koichi Narasaka 1255

34   **Additions to Enolates**

34.1 Alkylation of Enolates
      David L. Hughes 1273

34.2 Protonation of Enolates
      Akira Yanagisawa, Hisashi Yamamoto 1295

35   **Ring Opening of Epoxides and Related Reactions**
      Eric N. Jacobsen, Michael H. Wu. 1309

36   **Polymerization Reactions**
      Geoffrey W. Coates 1329

37   **Heterogeneous Catalysis**
      Hans-Ulrich Blaser, Martin Studer 1353

38   **Catalyst Immobilization**

38.1 Catalyst Immobilization: Solid Supports
      Benoît Pugin, Hans-Ulrich Blaser. 1367

38.2 Catalyst Immobilization: Two-Phase Systems
      Günther Oehme 1377

39   **Combinatorial Approaches**
      Ken D. Shimizu, Marc L. Snapper, Amir H. Hoveyda 1389

40   **Catalytic Antibodies**
      Paul Wentworth Jr., Kim D. Janda 1403

41   **Industrial Applications**

41.1 The Chiral Switch of Metolachlor
      Hans-Ulrich Blaser, Felix Spindler 1427

41.2 Process R&D of Pharmaceuticals, Vitamins,
      and Fine Chemicals
      Rudolf Schmid, Michelangelo Scalone 1439

41.3 Cyclopropanation
      Tadatoshi Aratani 1451

41.4 Asymmetric Isomerization of Olefins
      Susumu Akutagawa 1461

42   **Future Perspectives in Asymmetric Catalysis**
      Eric N. Jacobsen 1473

Subject Index 1479

Supplement to Chapter 5.2
Hydrogenation of Non-Functionalized Carbon–Carbon Double Bonds

Ronald L. Halterman

Department of Chemistry and Biochemistry, University of Oklahoma,
620 Parrington Oval, Norman, OK 73019, USA
e-mail: rhaltermann@ou.edu

Keywords: Asymmetric hydrogenation, Unfunctionalized alkenes, Cationic chiral zirconocene, Brintzinger complex, Crabtree hydrogenation catalysts, Chiral iridium complexes, Chiral oxazoline, Chiral phosphine, Chiral phosphinoethyloxazoline, Chiral phosphinite, Chiral imidazolylidine, *JM*-Phos

1   Introduction   1

2   Chiral Cationic Zirconocene Complexes.   2

3   Chiral Iridium Complexes   3

References   6

1
Introduction

The highly enantioselective hydrogenation of functionalized alkenes containing coordinating groups such as allylic alcohol, amide, or carbonyl functionalities has recently continued to be advanced primarily by using chiral phosphine and related donor ligand complexes of ruthenium and rhodium [1]. These complexes, however, have not had the same success in hydrogenating unfunctionalized alkenes containing only hydrocarbon substituents. Without the ability to achieve the prior coordination of the functional group, the isolated carbon–carbon double bond is likely to be too unreactive to hydrogenate efficiently with these less electrophilic ruthenium or rhodium catalysts. Two classes of enantioselective hydrogenation catalysts, the Crabtree-type complexes and the cationic Brintzinger-type zirconocene complexes, have become more established since the previous review [2] as the most effective for unfunctionalized alkenes are both highly electrophilic and thus more likely to coordinate and react with isolated alkene moieties. This chapter will update the progress with these two types of catalysts, including the first use of a chiral heterocyclic carbene as a phosphine replacement in enantioselective hydrogenations. The chapter concludes with new mechanistic studies involving deuterium labeling that point out limitations to judging the enantioselectivity of the alkene-catalyst interactions based solely on enantiomeric ratios of the hydrogenated products.

The hydrogenation of arylalkenes such as the substituted styrene shown below has in these recent reports been exclusively investigated. This choice of substrate is more determined by analytical problems than by reactivity issues [2]. Very often a 4-methoxyphenyl substituent is included as a "non-coordinating" aid for the chiral HPLC and GC enantiomeric purity measurements.

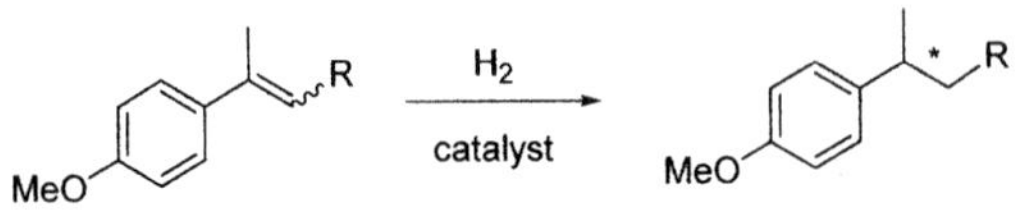

$$\tag{1}$$

2
Chiral Cationic Zirconocene Complexes

The hydrogenation of unfunctionalized trisubstituted alkenes with the reduced titanocene complex (S,S)-(ethylenebistetrahydroindenyl)titanium hydride [(EBTHI)TiH] was found to be very sluggish [3] and it was feared that this problem would be exacerbated when more hindered tetrasubstituted alkenes were investigated. Based in part on the high electrophilicity and reactivity of cationic zirconocenes [(EBTHI)ZrMe]$^+$ in the polymerization of alkenes, Buchwald instead investigated the use of (S,S)- or (R,R)-(EBTHI)ZrMe$_2$/[PhMe$_2$NH]$^+$ [(BC$_6$F$_5$)$_4$]$^-$ in the hydrogenation of such tetrasubstituted alkenes [4]. The catalytically active cationic species 1 is postulated to form in the presence of hydrogen and react as shown in Scheme 1. Indenes such as 2,3-dimethylindene react either at low pressure of hydrogen (80 psig) to give lower yields of the product 1,2-dimethylindane in 86% ee and a *cis/trans* ratio of 95:5 or at higher pressure (1,700 psig) to give an improved 93% ee and >99:1 *cis/trans* ratio. These improvements are probably due to the decreased likelihood of alkene isomerization via β-hydride elimination under higher pressure of hydrogen. Only in this case was the absolute configuration determined [(S,S)-product from (R,R)-catalyst]. A summary

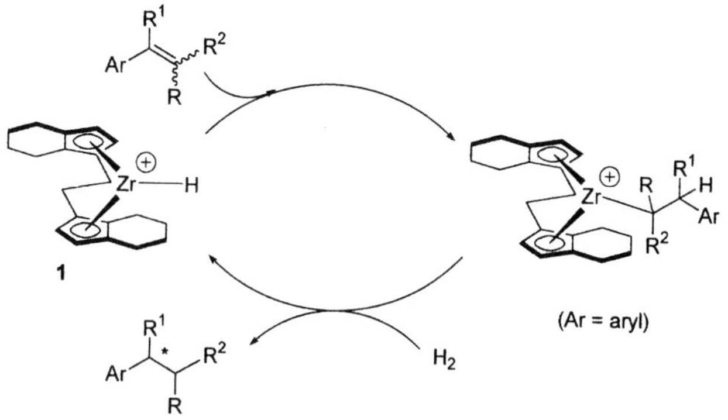

Scheme 1 Zirconocene-catalyzed hydrogenation

Table 1 Enantioselective hydrogenation of alkenes 2–4 with complex1[a]

Alkene	Catalyst loading (mol%)	H$_2$ pressure (psig)	Yield (%)	ee (%)	cis:trans
2a	5	80	76	86	95:5
2a	8	1,700	87	93	>99:1
2b	8	80	96	92	
2c	5	80	34[b]	97	
3a	5	1,000	80[b]	40	95:5
3b	8	2,000	94	78	>99:1
4	5	2,000	91	92	>99:1

[a] Reactions run at 0.25 M [alkene], room temperature, 13–21 h
[b] Percentage conversion

of results using **1** is given in Table 1. The phenyl-substituted substrates are noticeably lower in reactivity, although the selectivity is high. Even though the catalyst loading is high at 5–8 mol%, these catalysts are remarkable in their ability to reduce these tetrasubstituted alkenes.

2a (R = Me)   **3a** (R = Et)
2b (R = n-Bu) **3b** (R = Ph)   **4**
2c (R = Ph)

(2)

3
Chiral Iridium Complexes

Since Pfaltz's introduction of chiral phosphine–oxazoline (PHOX) [5] complexes of iridium such as **5** [6], the use of similar highly electrophilic Crabtree-type catalysts for the enantioselective hydrogenation of unfunctionalized alkenes has recently seen the most progress. The results of selected, representative complexes **5–10** for the hydrogenation of aryl-substituted alkenes **11–16** are presented in Table 2. Each complex has in common the incorporation of a chiral oxazoline moiety tethered to either an achiral phosphine (PHOX **5** [6], *JM*-Phos **9** [7]), phosphinopyrrolyl (Pyr-PHOX **6** [8]), phosphinite (serine-derived phosphinite–oxazoline **7** [9] and threonine-derived phosphinite–oxazoline **8** [10]), or imidazolylidine (imidazolylidine–oxazoline **10** [11]) moiety. The modular nature of the ligand preparation from various oxazoline precursors has enabled libraries of up to 10 derivatives of the complexes to be generated and studied [12]. Since representative results of only one or two derivatives of each complex under comparable reaction conditions are given here, higher selectivities

Table 2 Enantioselective hydrogenation of alkenes 11–16 with complexes 5–10

Alkene	Complex	mol%	$P(H_2)$, T, time (bar, °C, h)	Conversion (%)	% ee (configuration)
11	PHOX 5a	1.0	50, 23, 2	100	99 (R)
11	Pyr-PHOX 6	1.0	50, 23, 2	100	99 (R)
11	Serine–phosphinite– oxazoline 7	0.1	50, 23, 2	100	97 (R)
11	Threonine-phosphi- nite-oxazoline 8	0.02	50, 23, 4	100	99 (R)
11	JM-Phos 9b	0.2	50, 25, 2	99	95 (S)
11	Imidazolylidine–ox- azoline 10	0.2–0.6	50, 25, 2	98	99 (S)
12	JM-Phos 9b	0.3	50, 25, 2	99	93 (S)
12	Imidazolylidine–ox- azoline 10	0.6	50, 25, 2	99	97 (S)
13	PHOX 5b	1.0	50, 23, 2	100	72 (S)
13	Pyr-PHOX 6	1.0	50, 23, 2	100	90 (S)
13	Serine–phosphinite– oxazoline 7	0.5	50, 23, 2	100	85 (S)
13	Threonine–phosphi- nite–oxazoline 8	1.0	50, 23, 2	100	71 (S)
14	Serine–phosphinite– oxazoline 7	0.1	1, 23, 2	100	78 (S)
14	Threonine–phosphi- nite–oxazoline 8	1.0	50, 23, 0.5	100	62 (R)
14	Threonine–phosphi- nite–oxazoline 8	1.0	1, 0, 0.5	100	89 (R)
14	JM-Phos 9a	0.3	50, 60, 2	99	44 (R)
14	Imidazolylidine–ox- azoline 10	0.3	50, 25, 2	91	31 (R)
E-15	PHOX 5b	1.0	50, 23, 2	100	81 (R)
E-15	Pyr-PHOX 6	1.0	50, 23, 2	100	75 (R)
E-15	Serine–phosphinite– oxazoline 7	1.1	50, 23, 2	100	93 (R)
E-15	Threonine–phosphi- nite-oxazoline 8	1.0	50, 23, 2	100	99 (R)
E-15	JM-Phos 9a	0.6	50, 25, 2	90	80 (S)
E-15	Imidazolylidine–ox- azoline 10	0.6	50, 25, 2	99	91 (S)
Z-16	PHOX 5b	1.0	50, 23, 2	100	63 (S)
Z-16	Pyr-PHOX 6	1.0	50, 23, 2	100	70 (S)
Z-16	Serine–phosphinite– oxazoline 7	0.4	50, 23, 2	100	85 (S)
Z-16	Threonine–phosphi- nite–oxazoline 8	1.0	50, 23, 2	100	89 (S)
Z-16	JM-Phos 9a	0.6	50, 25, 2	70	75 (R)
Z-16	Imidazolylidine–ox- azoline 10	1.0	50, 25, 2	95	78 (R)

and lower catalyst loadings can often be found in the original literature. In general these iridium complexes (0.1–1.0 mol%) will completely convert alkene to alkane within 2 h at room temperature at 0.3–1.0 M [alkene] in methylene chloride under 50 bar hydrogen pressure.

(3)

As can be seen from the table, the two methylstilbene derivatives **11** and **12** are reduced highly enantioselectively by each class of catalyst. The dihydronaphthylene derivative **13**, however, gave only selectivity up to 90% ee. In the series of substituted styrenes **14–16** which give the same (1-methylpropyl)aryl hydrogenation product, enantioselectivities were quite variable. With each catalyst, the reduction of *E*-**15** gave the enantiomer of the product obtained by hydrogenating the isomeric *Z*-**16** alkene. The catalysts can be grouped according to the position of the stereocenter in the oxazoline ring. In the PHOX-**5** and Pyr-PHOX-**6** complexes the tether to the phosphorus ligand is on a non-stereogenic oxazoline site, whereas in the remaining complexes, the attachment of the tethered donor ligand is at a stereocenter; these latter complexes are consistently more enantioselective in the hydrogenation of substituted styrenes *E*-**15** and *Z*-**16**.

An explanation for the relatively poor showing of some of these catalysts in the hydrogenation of substituted styrenes **14–16** versus the substituted stilbenes **11** and **12** has recently been proposed to be at least partially due to competing double bond migration. Burgess used deuteration experiments to show that deuterium is incorporated not only across the double bond, but also at the allylic position in the ratios given in Scheme 2 for two catalysts [7, 11]. Since the *E*-**15** and *Z*-**16** substrates preferentially lead to opposite enantiomers, any isomerization between these substrates will certainly lead to poorer observed enantioselectivities, even if the individual alkene–catalyst interactions are highly stereoselective. Interestingly, the use of an imidazolydinine donor ligand in **10** gave significantly less incorporation of deuterium at the allylic sites.

In summary, the range of highly electrophilic catalysts for the enantioselective hydrogenation of unfunctionalized trisubstituted and tetrasubstituted

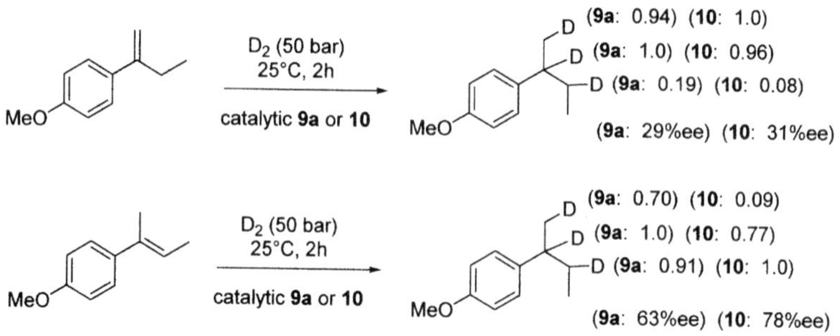

Scheme 2 Deuterium incorporation during hydrogenation

alkenes has been extended. With established reaction conditions and analytical techniques, a high level of enantioselectivity has been achieved.

References

1. Brown JM (1999) Hydrogenation of functionalized carbon-carbon double bonds. In: Jacobsen EN, Pfaltz A, Yamamota H (eds) Comprehensive Asymmetric Catalysis. Springer, Berlin Heidelberg New York, p 122
2. Halterman RL (1999) Hydrogenation of non-functionalized carbon-carbon double bonds. In: Jacobsen EN, Pfaltz A, Yamamota H (eds) Comprehensive asymmetric catalysis. Springer, Berlin Heidelberg New York, p 183
3. Broene RD, Buchwald SL (1993) J Am Chem Soc 115:12569
4. Troutman MV, Appella DH, Buchwald SL (1999) J Am Chem Soc 121:4916
5. Helmchen G, Pfaltz A (2000) Acc Chem Res 33:336
6. Blackmond DG, Lightfoot A, Pfaltz A, Rosner T, Schider P, Zimmermann N (2000) Chirality 12:442
7. Hou D-R, Reibenspies J, Colacot TJ, Burgess K (2001) Chem Eur J 7:5391
8. Cozzi PG, Zimmermann N, Hilgraf R, Schaffner S, Pfaltz A (2001) Adv Syn Catal 343:450
9. Blankenstein J, Pfaltz A (2001) Angew Chem 113:4577; Angew Chem Int Ed Engl 40:4445
10. Menges F, Pfaltz A (2002) Adv Syn Catal 344:40
11. Powell MT, Hou D-R, Perry MC, Cui X, Burgess K (2001) J Am Chem Soc 123:8878
12. Pfaltz A, Blankenstein J, Hilgraf R, Hörmann E, McIntyre S, Menges F, Schönleber M, Smidt SP, Wüstenberg B, Zimmermann N (2003) Adv Syn Catal 345 (in press)

Supplement to Chapter 6.4
Hydroboration of Carbonyl Groups

Takeshi Ohkuma, Ryoji Noyori

Department of Chemistry and Research Center for Materials Science, Nagoya University, Chikusa, 464-8602, Nagoya, Japan
e-mail: noyori@chem3.chem.nagoya-u.ac.jp

Keywords: Hydroboration, Ketone, Borane reagents, Borohydride

1	**Hydroboration Catalyzed by Oxazaborolidines and Their Analogues**	7
1.1	Borane Reagents	8
1.2	Reduction of α-Hetero Substituted Aliphatic Ketones	8
1.3	Variation of the Catalyst	8
1.4	Immobilized Catalysts	10
2	**Hydroboration Catalyzed by Metal Complexes**	12
3	**Chemoselective and Stereoselective Reduction with NaBH4 Catalyzed by Co Complexes**	14
References	. . . .	18

1
Hydroboration Catalyzed by Oxazaborolidines and Their Analogues

The pioneering studies by Itsuno [1] and Corey [2] on the development of the asymmetric hydroboration of ketones using oxazaborolidines have made it possible to easily obtain chiral secondary alcohols with excellent optical purity [3]. Scheme 1 shows examples of Corey's (Corey–Bakshi–Shibata) CBS reduction. When oxazaborolidines 1 were used as catalysts (usually 0.01–0.1 equiv), a wide variety of ketones were reduced by borane reagents with consistently high enantioselectivity [2]. The sense of enantioselection was predictable. Many important biologically active compounds and functional materials have been synthesized using this versatile reaction [2–4].

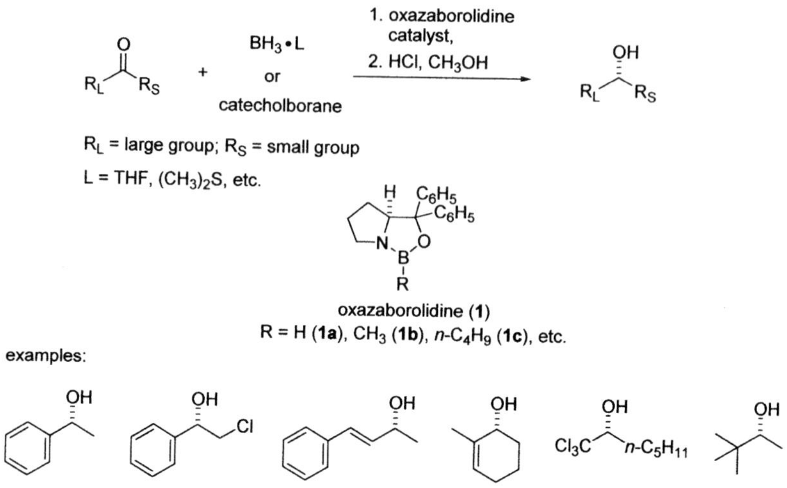

Scheme 1

1.1
Borane Reagents

Corey originally used borane–THF as a stoichiometric reducing agent [2, 5]. The use of more robust borane sources, borane-N,N-diethylaniline [6, 7] and borane-N-ethyl-N-isopropylaniline [8, 9], rendered the CBS reduction easier to handle without sacrificing enantioselectivity. Borane–THF prepared in situ from NaBH$_4$ and (CH$_3$)$_3$SiCl in THF [10] was also shown to be useful for this type of reduction [11].

1.2
Reduction of α-Hetero Substituted Aliphatic Ketones

Extensive studies by Corey have clearly revealed the wide range of applications of the CBS reduction [2]. However, recent observations of high enantioselectivity in the reduction of several α-hetero substituted aliphatic ketones catalyzed by **1** are worthy of mention (Scheme 2). A series of aliphatic α-hydroxyketones protected with tetrahydropyranyl [7], trialkylsilyl [12], and sulfonyl groups [8] as well as β-ketosulfones [9] were reduced to give the corresponding alcohols in up to >99% ee.

1.3
Variation of the Catalyst

B-Methoxy oxazaborolidine **2** was prepared from the corresponding amino alcohol and trimethyl borate (Scheme 3) [13, 14]. The Lewis acidity, which is high-

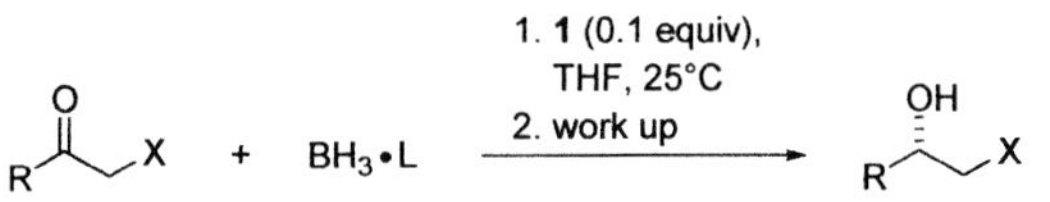

R	X	L	1	Yield (%)	ee (%)
C_2H_5	$OTHP^a$	N,N-diethylaniline	1b	92	86
$(CH_3)_2CH$	$OTHP^a$	N,N-diethylaniline	1b	92	91
cyclo-C_6H_{11}	$OTHP^a$	N,N-diethylaniline	1b	93	97
$(CH_3)_3C$	$OTHP^a$	N,N-diethylaniline	1b	92	97
C_2H_5	$OSi(C_2H_5)_3$	THF	1a	82^b	73
cyclo-C_6H_{11}	$OSi(C_2H_5)_3$	THF	1a	84^b	96
cyclo-C_6H_{11}	OSO_2CH_3	N-ethyl-N-isopropylaniline	1b	96	96
n-$C_{11}H_{23}$	$SO_2C_6H_4$-p-CH_3	N-ethyl-N-isopropylaniline	1b	97	87
cyclo-C_6H_{11}	$SO_2C_6H_4$-p-CH_3	N-ethyl-N-isopropylaniline	1b	99	>99
$(CH_3)_3C$	$SO_2C_6H_4$-p-CH_3	N-ethyl-N-isopropylaniline	1b	98	99

Scheme 2

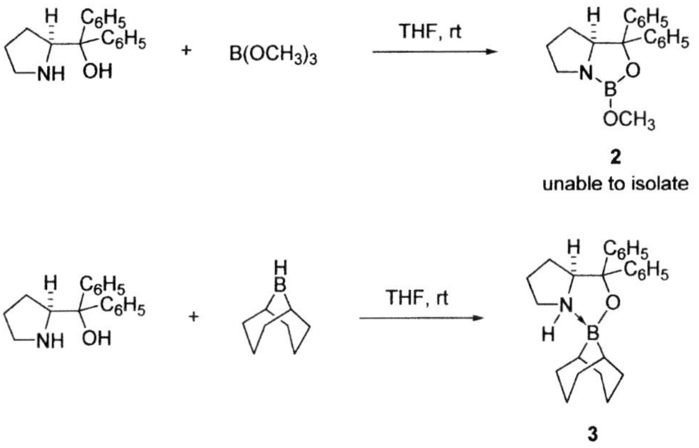

Scheme 3

er than that of **1b**, was expected to increase both reactivity and enantioselectivity in the reduction of ketonic substrates. Optical yields of up to 99% in the reduction of acetyl pyridines with **2** were higher than those in the reaction catalyzed by **1b** [13].

Preparation of **1b** from the amino alcohol and methaneboronic acid is required to completely remove water [2]. Catalyst **3** was more conveniently prepared from the same amino alcohol and 9-borabicyclo[3.3.1]nonane, accompanied by the release of hydrogen instead of water (Scheme 3) [15]. Catalyst **3** resulted in an excellent optical yield of up to 99.2% in the hydroboration of relatively hindered aromatic ketones.

The excellent enantioselectivity and wide scope of the CBS reduction have motivated researchers to make new chiral auxiliaries [3]. Figure 1 depicts examples of in situ prepared and preformed catalyst systems reported since 1997. Most of these amino-alcohol-derived catalysts were used for the reduction of α-halogenated ketones and/or for the double reduction of diketones [16–28]. Sulfonamides [29, 30], phosphinamides [31], phosphoramides [32], and amine oxides [33] derived from chiral amino alcohols were also applied. The reduction of aromatic ketones with a chiral 1,2-diamine [34] and an α-hydroxythiol [35] gave good optical yields. Acetophenone was reduced with borane–THF in the presence of a chiral phosphoramidite with an optical yield of 96% [36].

1.4
Immobilized Catalysts

As regards separation from products and recycling, heterogeneous catalysts are clearly advantageous over homogeneous catalysts. Several attempts to immobilize oxazaborolidines on polymers have been reported [3]. Recently, a catalyst system prepared from a polymer-bound chiral sulfonamide and borane–dimethyl sulfide **4** was applied to the hydroboration of aromatic ketones (Fig. 2) [37]; an optical yield of up to 96% was achieved. The enantioselectivity was higher than that observed in the reduction with the corresponding homogeneous catalyst [29]. The catalyst was recovered by filtration and could be reused five times without a loss of enantioselectivity. Borane prepared in situ from $NaBH_4$ and $(CH_3)_3SiCl$ [10] or BF_3–ether [38] was also used for this purpose with the same immobilized catalyst [39].

In reactions with polymer-bound catalysts, a mass-transfer limitation often results in slowing down the rate of the reaction. To avoid this disadvantage, homogenous organic-soluble polymers have been utilized as catalyst supports. Oxazaborolidine **5**, supported on linear polystyrene, was used as a soluble immobilized catalyst for the hydroboration of aromatic ketones in THF to afford chiral alcohols with an ee of up to 99% [40]. The catalyst was separated from the products with a nanofiltration membrane and then was used repeatedly. The total turnover number of the catalyst reached as high as 560. An intramolecularly cross-linked polymer molecule (microgel) was also applicable as a soluble support [41].

Nickel boride prepared from NiI_2 and two equivalents of $LiBH_4$ [42] was utilized as an oxazaborolidine catalyst support (Scheme 4) [43]. Reaction of nickel boride with 0.1 equivalents of chiral amino alcohol in THF at room temperature gave the anchored catalyst **6**, which produced chiral alcohols in optical yields of up to 95%, and which furthermore showed higher activity as regards the reduction of acetophenone derivatives than that of the corresponding homogene-

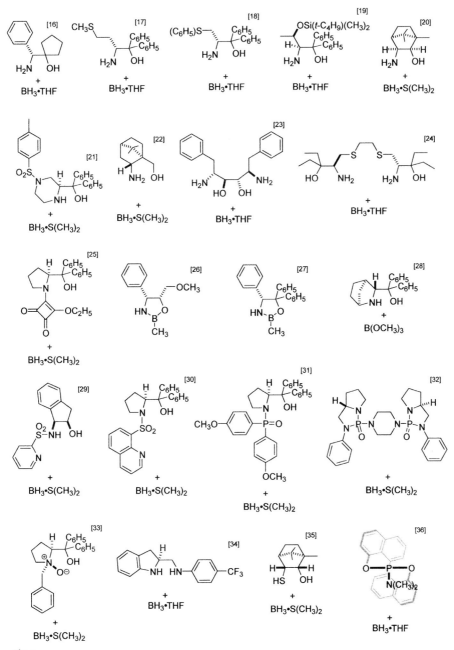

Fig. 1

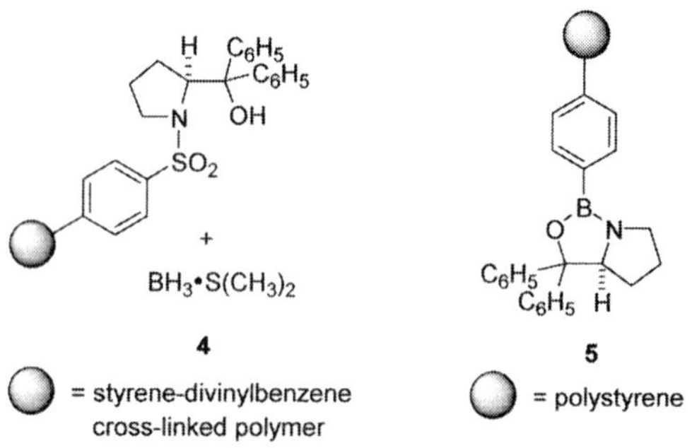

Fig. 2

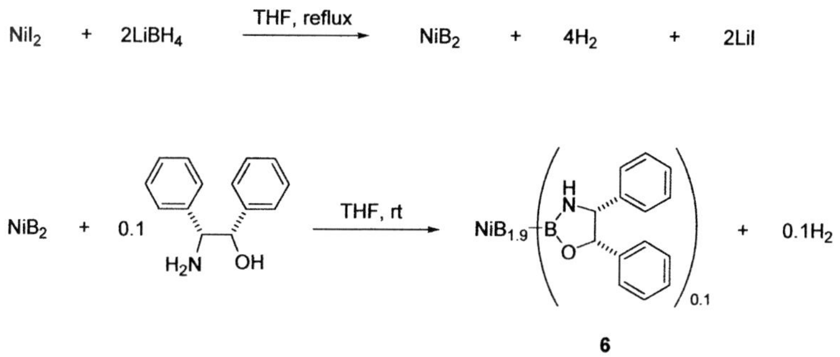

Scheme 4

ous catalyst. Anchored catalyst **6** was reusable three times without compromising efficiency.

2
Hydroboration Catalyzed by Metal Complexes

Ti complexes prepared from Ti[OCH(CH$_3$)$_2$]$_4$ and one equivalent of chiral diol catalyzed the reduction of ketones with catecholborane [44]. Scheme 5 shows a recent example utilizing anisyl-BODOL (**7**) as a chiral auxiliary [45]. Several aromatic ketones and 2-acetylcyclohexanone were reduced with an optical yield of up to 98%. Reduction of 2-octanone, a simple aliphatic ketone, gave the corresponding alcohol in 87% ee.

Asymmetric hydroboration of ketones with catecholborane in the presence of 0.02–0.025 equivalents of (*R*,*R*)-**8**, which was prepared from LiGaH$_4$ and 2 equivalents of 2-hydroxy-2′-mercapto-1,1′-binaphthyl, afforded the corresponding alcohols in a high yield and with high enantioselectivity (Scheme 6) [46]. The solid-state structure of **8** is drawn in this scheme. The preferential interaction between soft Ga metal and soft thiolate is supposed to prevent the replacement of the chiral ligand by the alkoxide product.

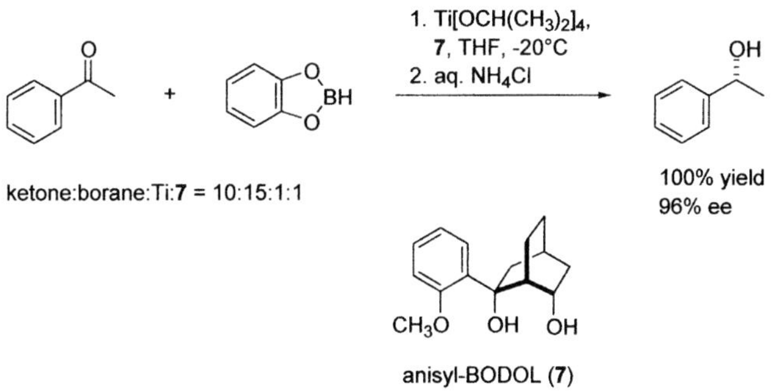

ketone:borane:Ti:**7** = 10:15:1:1

1. Ti[OCH(CH₃)₂]₄,
   7, THF, -20°C
2. aq. NH₄Cl

100% yield
96% ee

CH₃O   OH   OH

anisyl-BODOL (**7**)

Scheme 5

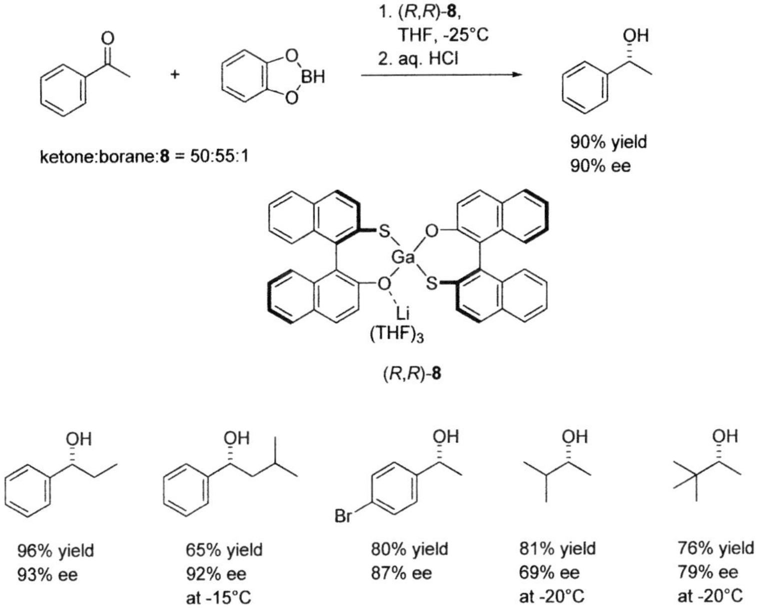

ketone:borane:**8** = 50:55:1

1. (R,R)-**8**,
   THF, -25°C
2. aq. HCl

90% yield
90% ee

(THF)₃

(R,R)-**8**

OH	OH	OH	OH	OH
96% yield	65% yield	80% yield	81% yield	76% yield
93% ee	92% ee	87% ee	69% ee	79% ee
	at -15°C		at -20°C	at -20°C

Scheme 6

Al complexes prepared in situ from Al[OCH(CH₃)₂]₃ and two equivalents of (R)-BINAPHTHOL (**9**) and (R)-H₈-BINAPHTHOL (**10**) promoted the enantioselective reduction of propiophenone with borane–dimethyl sulfide and gave the S alcohol in 83% and 90% ee, respectively (Scheme 7) [47]. The reaction was much slower and afforded a racemic product in the absence of Al[OCH(CH₃)₂]₃ under otherwise identical conditions. The addition of a catalytic amount of Al(OC₂H₅)₃ increased both the rate and enantioselectivity in the hydroboration of ketones with a chiral amino alcohol [48].

The chiral BINAPHTHOL derivative (R)-**11** and two equivalents of Zn(C₂H₅)₂ produced an active catalyst for the asymmetric reduction of acetophenone with

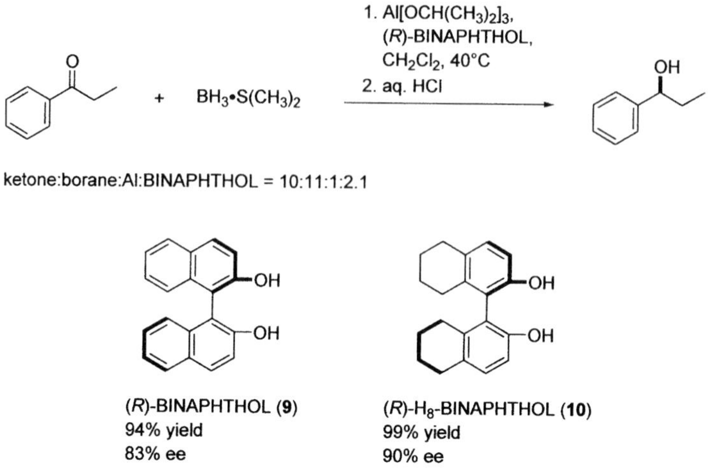

1. Al[OCH(CH$_3$)$_2$]$_3$,
   (R)-BINAPHTHOL,
   CH$_2$Cl$_2$, 40°C
2. aq. HCl

ketone:borane:Al:BINAPHTHOL = 10:11:1:2.1

(R)-BINAPHTHOL (9)
94% yield
83% ee

(R)-H$_8$-BINAPHTHOL (10)
99% yield
90% ee

Scheme 7

catecholborane that gave (S)-1-phenylethanol in 81% ee (Fig. 3) [49]. The polymeric catalyst prepared from (R)-**12** and Zn(C$_2$H$_5$)$_2$ produced the same enantioselectivity [49]. α-Methoxyacetophenone was reduced with catecholborane in the presence of a catalyst prepared from chiral bisoxazoline (R,R)-**13** and an equimolar amount of Zn(OSO$_2$CF$_3$)$_2$ to afford the S alcohol in 82% ee (Fig. 3) [50].

3
Chemoselective and Stereoselective Reduction with NaBH$_4$ Catalyzed by Co Complexes

NaBH$_4$ is one of the most versatile and widely used metal hydride reagents [51]. Mukaiyama has devised a highly enantioselective reduction of simple aromatic ketones using NaBH$_4$ as a stoichiometric reducing agent catalyzed by β-keto iminato Co complexes [52]. For example, butyrophenone was reduced essentially quantitatively with (S,S)-**14** in the presence of tetrahydrofurfuryl alcohol (THFA) and ethanol in CHCl$_3$ at -20°C to give the S alcohol in 97% ee (Scheme 8). A modified borohydride NaBH$_2$(OR)(OC$_2$H$_5$) (ROH=THFA) was supposed to be a stoichiometric reducing agent [53].

Recently, a series of studies of the stereoselective reduction of 1,3-diketones has been reported which expands upon this methodology. When 1,3-diphenyl-1,3-propanedione was reduced with three equivalents of NaBH$_4$ in the presence of (S,S)-**15**, THFA, and ethanol in CHCl$_3$ at -20°C, (S,S)-1,3-diphenyl-1,3-propanediol was obtained quantitatively in 97% ee (Scheme 9) [54]. The *anti:syn* ratio of the product was 84:16. Diketones with electron-donating and electron-withdrawing substituents on their phenyl rings were also reduced with high enantioselectivity and diastereoselectivity.

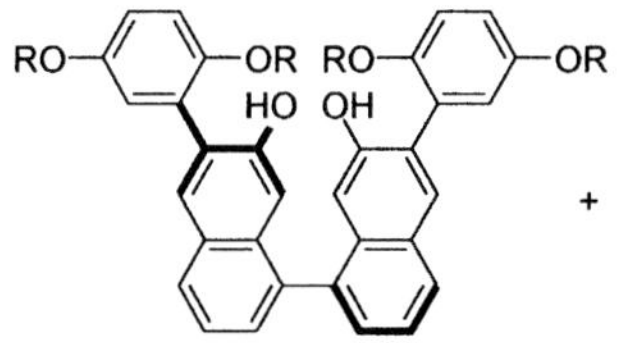

(R)-11: R = n-C$_6$H$_{13}$

+   Zn(C$_2$H$_5$)$_2$ (2 equiv)

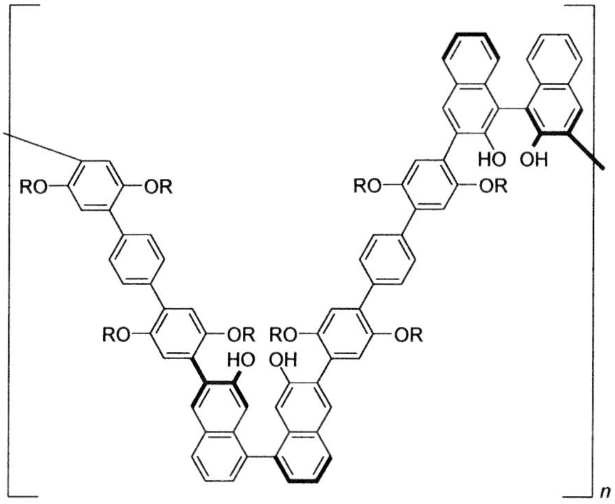

+   Zn(C$_2$H$_5$)$_2$

(2 equiv for each
binaphthyl unit)

(R)-12: R = n-C$_6$H$_{13}$

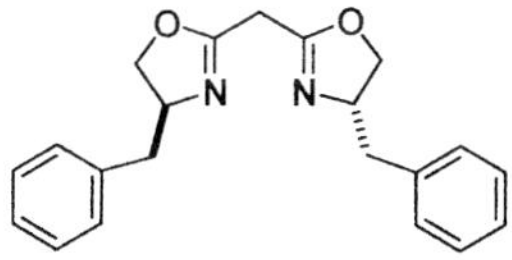

+   Zn(OSO$_2$CF$_3$)$_2$

(S,S)-13

Fig. 3

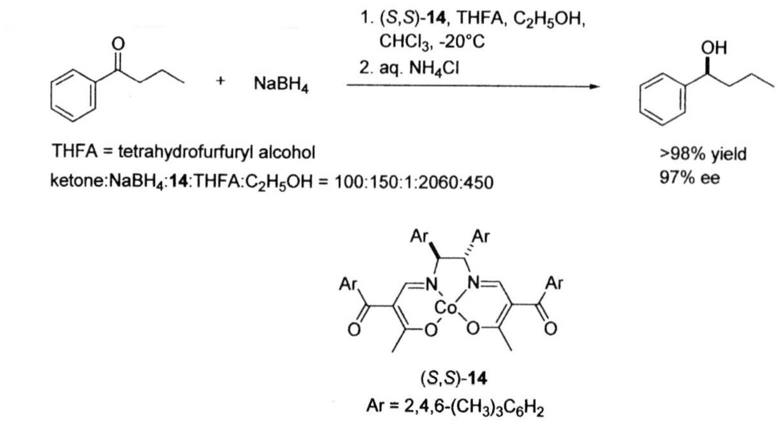

1. (S,S)-**14**, THFA, C_2H_5OH,
   $CHCl_3$, -20°C
2. aq. NH_4Cl

THFA = tetrahydrofurfuryl alcohol
ketone:NaBH$_4$:**14**:THFA:C_2H_5OH = 100:150:1:2060:450

>98% yield
97% ee

(S,S)-**14**
Ar = 2,4,6-$(CH_3)_3C_6H_2$

Scheme 8

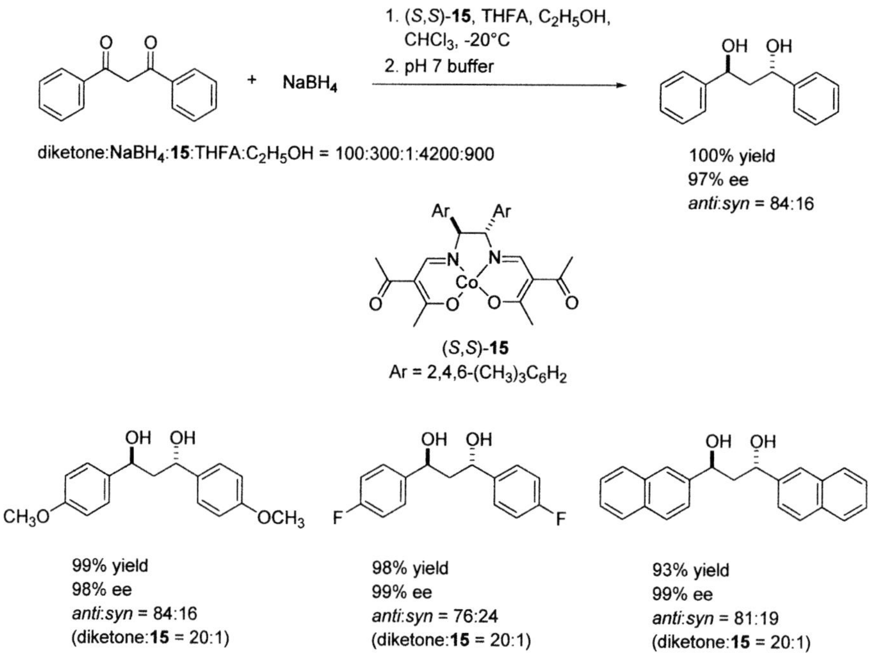

1. (S,S)-**15**, THFA, C_2H_5OH,
   $CHCl_3$, -20°C
2. pH 7 buffer

diketone:NaBH$_4$:**15**:THFA:C_2H_5OH = 100:300:1:4200:900

100% yield
97% ee
anti:syn = 84:16

(S,S)-**15**
Ar = 2,4,6-$(CH_3)_3C_6H_2$

99% yield
98% ee
anti:syn = 84:16
(diketone:**15** = 20:1)

98% yield
99% ee
anti:syn = 76:24
(diketone:**15** = 20:1)

93% yield
99% ee
anti:syn = 81:19
(diketone:**15** = 20:1)

Scheme 9

The desymmetrization of 2-alkyl-1,3-diketones to the corresponding chiral hydroxyketones was also successfully achieved with the same catalyst system. For example, 2-methyl-1,3-diphenyl-1,3-propanedione was reduced with an equimolar amount of NaBH$_4$ together with THFA and ethanol in the presence of 0.05 equivalents of (R,R)-**15** to afford (1R,2S)-2-methyl-3-oxo-1,3-diphenyl-propane (anti:syn=99:1) in 99% ee (Scheme 10) [55]. 2-Allyl- and 2-benzyl-sub-

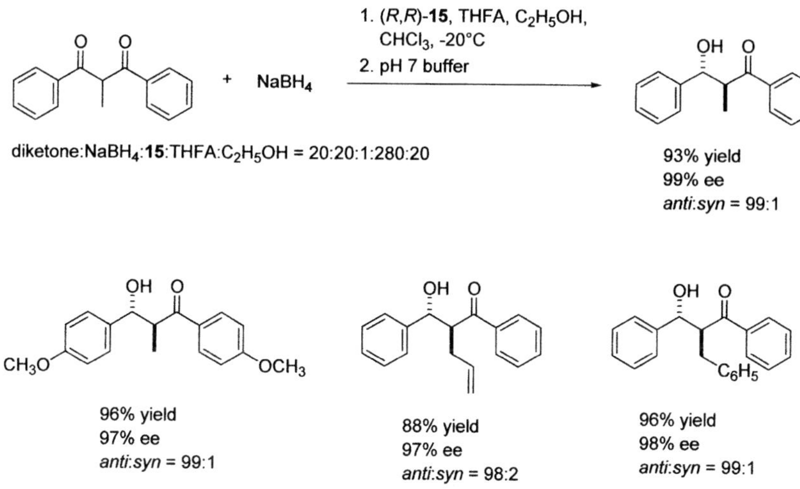
1. (R,R)-15, THFA, C2H5OH, CHCl3, -20°C
2. pH 7 buffer
+ NaBH4
diketone:NaBH4:15:THFA:C2H5OH = 20:20:1:280:20
93% yield
99% ee
anti:syn = 99:1
CH3O
OCH3
96% yield
97% ee
anti:syn = 99:1
88% yield
97% ee
anti:syn = 98:2
C6H5
96% yield
98% ee
anti:syn = 99:1

Scheme 10

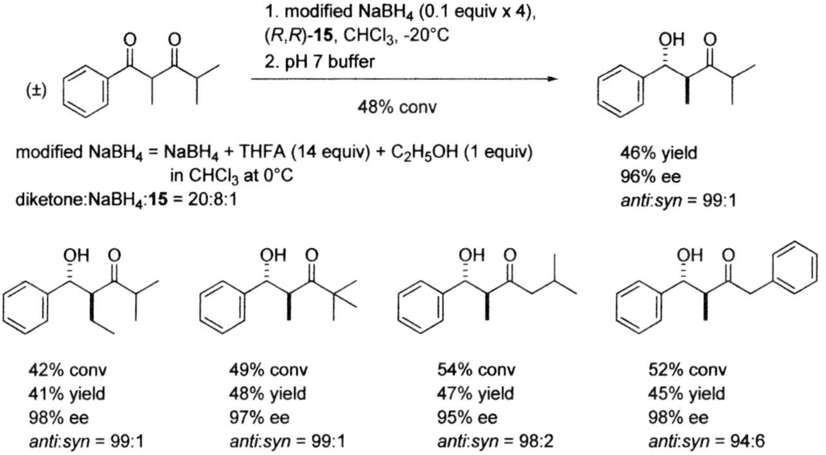
1. modified NaBH4 (0.1 equiv x 4), (R,R)-15, CHCl3, -20°C
2. pH 7 buffer
48% conv
(±)
modified NaBH4 = NaBH4 + THFA (14 equiv) + C2H5OH (1 equiv) in CHCl3 at 0°C
diketone:NaBH4:15 = 20:8:1
46% yield
96% ee
anti:syn = 99:1
42% conv
41% yield
98% ee
anti:syn = 99:1
49% conv
48% yield
97% ee
anti:syn = 99:1
54% conv
47% yield
95% ee
anti:syn = 98:2
52% conv
45% yield
98% ee
anti:syn = 94:6

Scheme 11

stituted diketones were reduced with a similarly high level of stereoselectivity. The high degree of diastereoselectivity was interpreted by applying the Felkin–Anh rule [56].

Finally, this catalyst system was applied to the chemoselective, diastereoselective, and enantioselective reduction of racemic 2-alkyl-1,3-diketones [57]. Scheme 11 shows examples of this transformation catalyzed by (*R*,*R*)-**15**. To minimize the effects of uncatalyzed reduction, four portions of 0.1 equivalents of modified borohydride with THFA and ethanol (i.e., total 0.4 equiv) were successively added to the substrate solution at –20°C. The selectivity of the carbonyl group at the benzylic position over the simple aliphatic carbonyl function (in a ratio of up to 99:1) was much higher than that observed in the reduction with the original $NaBH_4$. The optical yield of >95% at close to 50% conversion was clearly the result of the high enantiomer-selective ability of the catalyst.

References

1. Itsuno S, Nakano M, Miyazaki K, Masuda H, Ito K, Hirao A, Nakahama S (1985) J Chem Soc Perkin 1 2039; Itsuno S, Ito K, Maruyama T, Kanda N, Hirao A, Nakahama S (1986) Bull Chem Soc Jpn 59:3329
2. Corey EJ, Helal CJ (1998) Angew Chem Int Ed 37:1986
3. Recent reviews: Itsuno S (1998) Org React 52:395; Itsuno S (1999) Hydroboration of carbonyl groups. In: Jacobsen EN, Pfaltz A, Hamamoto H (eds) Comprehensive asymmetric catalysis, vol 1, Chap 6.4. Springer, Berlin Heidelberg New York; Carboni B, Monnier L (1999) Tetrahedron 55:1197; Cho BT (2002) Aldrichimica Acta 35.3
4. See for example: Hett B, Fang QK, Gao Y, Hong Y, Butler HT, Nie X, Wald SA (1997) Tetrahedron Lett 38:1125; Lotz M, Ireland T, Tappe K, Knochel P (2000) Chirality 12:389; Fehr C, Galindo J (2000) Angew Chem Int Ed 39:569; Mellin-Morlière C, Aitken DJ, Bull SD, Davies SG, Husson H-P (2001) Tetrahedron Asymmetry 12:149; Albrecht M, Kocks BM, Spek AL, van Koten G (2001) J Organomet Chem 624:271; Choi OK, Cho BT (2001) Tetrahedron Asymmetry 12:903
5. Corey EJ, Bakshi RK, Shibata S (1987) J Am Chem Soc 109:5551
6. Salunkhe AM, Burkhardt ER (1997) Tetrahedron Lett 38:1523; Cho BT, Chun YS (1999) J Chem Soc Perkin 1:2095
7. Cho BT, Chun YS (1999) Tetrahedron Asymmetry 10:1843
8. Cho BT, Yang WK, Choi OK (2001) J Chem Soc Perkin 1 1204
9. Cho BT, Kim DJ (2001) Tetrahedron Asymmetry 12:2043
10. Giannis A, Sandhoff K (1989) Angew Chem Int Ed Engl 28:218; Bolm C, Seger A, Felder M (1993) Tetrahedron Lett 34:8079
11. Jiang B, Feng Y, Zheng J (2000) Tetrahedron Lett 41:10281
12. Cho BT, Chun YS (1998) J Org Chem 63:5280
13. Masui M, Shioiri T (1997) Synlett 273
14. Masui M, Shioiri T (1998) Tetrahedron Lett 39:5195
15. Kanth JVB, Brown HC (2002) Tetrahedron 58:1069
16. Reiners I, Martens J (1997) Tetrahedron Asymmetry 8:27
17. Trentmann W, Mehler T, Martens J (1997) Tetrahedron Asymmetry 8:2033
18. Li X, Xie R (1997) Tetrahedron Asymmetry 8:2283
19. Shimizu M, Tsukamoto K, Matsutani T, Fujisawa T (1998) Tetrahedron 54:10265
20. Santhi V, Rao JM (2000) Tetrahedron Asymmetry 11:3553
21. Inoue T, Saito D, Komura K, Itsuno S (1999) Tetrahedron Lett 40:5379
22. Li X, Yeung C, Chan ASC, Yang T-K (1999) Tetrahedron Asymmetry 10:759
23. Jiang B, Feng Y, Hang J-F (2001) Tetrahedron Asymmetry 12:2323
24. Kossenjans M, Martens J (1998) Tetrahedron Asymmetry 9:1409
25. Zhou H-B, Zhang J, Lü S-M, Xie R-G, Zhou Z-Y, Choi MCK, Chan ASC, Yang T-K (2001) Tetrahedron 57:9325
26. Puigjaner C, Vidal-Ferran A, Moyano A, Pericàs MA, Riera A (1999) J Org Chem 64:7902
27. Bach J, Berenguer R, Garcia J, Loscertales T, Manzanal J, Vilarrasa J (1997) Tetrahedron Lett 38:1091
28. Pinho P, Guijarro D, Andersson PG (1998) Tetrahedron 54:7897
29. Sibi MP, Cook GR, Liu P (1999) Tetrahedron Lett 40:2477
30. Yang G-S, Hu J-B, Zhao G, Ding Y, Tang M-H (1999) Tetrahedron Asymmetry 10:4307
31. Gamble MP, Smith ARC, Wills M (1998) J Org Chem 63:6068; Wills M, Gamble M, Palmer M, Smith A, Studley J, Kenny J (1999) J Mol Catal A Chemical 146:139
32. Basavaiah D, Reddy GJ, Chandrashekar V (2001) Tetrahedron Asymmetry 12:685

33. O'Neil IA, Turner CD, Kalindjian SB (1997) Synlett 777
34. Sato S, Watanabe H, Asami M (2000) Tetrahedron Asymmetry 11:4329
35. Yang T-K, Lee D-S (1999) Tetrahedron Asymmetry 10:405
36. Müller P, Nury P, Bernardinelli G (2000) Helv Chim Acta 83:843
37. Hu J, Zhao G, Yang G, Ding Z (2001) J Org Chem 66:303
38. Sengupta S, Sahu DP, Chatterjee SK (1994) Ind J Chem 33B:285
39. Hu J, Zhao G, Ding Z (2001) Angew Chem Int Ed 40:1109
40. Giffels G, Beliczey J, Felder M, Kragl U (1998) Tetrahedron Asymmetry 9:691
41. Schunicht C, Biffis A, Wulff G (2000) Tetrahedron 56:1693
42. Brown HC, Brown CA (1962) J Am Chem Soc 84:1495; Brown HC, Brown CA (1962) J Am Chem Soc 84:2827; Brown HC, Brown CA (1963) J Am Chem Soc 85:1003
43. Molvinger K, Lopez M, Court J (1999) J Mol Catal A Chemical 150:267; Molvinger K, Lopez M, Court J (1999) Tetrahedron Lett 40:8375; Molvinger K, Lopez M, Court J (2000) Tetrahedron Asymmetry 11:2263
44. Lindsley CW, DiMare M (1994) Tetrahedron Lett 35:5141; Giffels G, Dreisbach C, Kragl U, Weigerding M, Waldmann H, Wandrey C (1995) Angew Chem Int Ed Engl 34:2005
45. Almqvist F, Torstensson L, Gudmundsson A, Frejd T (1997) Angew Chem Int Ed Engl 36:376; Sarvary I, Almqvist F, Frejd T (2001) Chem Eur J 7:2158
46. Ford A, Woodward S (1999) Angew Chem Int Ed 38:335; Blake AJ, Cunningham A, Ford A, Teat SJ, Woodward S (2000) Chem Eur J 6:3586
47. Fu I-P, Uang B-J (2001) Tetrahedron Asymmetry 12:45; Lin Y-M, Fu I-P, Uang B-J (2001) Tetrahedron Asymmetry 12:3217
48. Yanagi T, Kikuchi K, Takeuchi H, Ishikawa T, Nishimura T, Kamijo T (1999) Chem Lett 1203
49. Huang W-S, Hu Q-S, Pu L (1999) J Org Chem 64:7940
50. Bandini M, Cozzi PG, de Angelis M, Umani-Ronchi A (2000) Tetrahedron Lett 41:1601
51. See for example: Brown HC, Ramachandran PV (1996) Sixty years of hydride reductions. In: Abdel-Magid AF (ed) Reduction in organic synthesis. American Chemical Society, Washington DC
52. Nagata T, Sugi KD, Yorozu K, Yamada T, Mukaiyama T (1998) Catal Surv Jpn 2:47; Yamada T, Nagata T, Ikeno T, Ohtsuka Y, Sagara A, Mukaiyama T (1999) Inorg Chim Acta 296:86
53. Nagata T, Sugi KD, Yamada T, Mukaiyama T (1996) Synlett 1076
54. Ohtsuka Y, Kubota T, Ikeno T, Nagata T, Yamada T (2000) Synlett 535
55. Ohtsuka Y, Koyasu K, Ikeno T, Yamada T (2001) Org Lett 3:2543
56. Chérest M, Felkin H, Prudent N (1968) Tetrahedron Lett 2199; Anh NT, Eisenstein O (1976) Tetrahedron Lett 155; Anh NT, Eisenstein O (1977) Nouv J Chim 1:61
57. Ohtsuka Y, Koyasu K, Miyazaki D, Ikeno T, Yamada T (2001) Org Lett 3:3421

Supplement to Chapter 20.1
Dihydroxylation of Carbon–Carbon Double Bonds

Annette Bayer

Department of Chemistry, University of Tromsø
9037, Tromsø, Norway
e-mail: annette.bayer@chem.uit.no

Keywords: Asymmetric dihydroxylation, Osmium tetroxide, Cinchona alkaloid, Ligand-accelerated catalysis, Immobilization

1	New Generations of Chiral Ligands	22
1.1	Substrate Scope and Limitations	22
1.2	Chiral Ligands for an Enantioselective Second Cycle	24
2	Improved Reaction Conditions	26
3	New Cooxidants	27
4	Immobilization of Osmium	30
5	Immobilization of the Chiral Ligand	34
5.1	Insoluble Organic Polymers	35
5.2	Inorganic Supports	37
5.3	Soluble Organic Polymers	38
6	Conclusion	41
	References	41

Abbreviations

AD: asymmetric dihydroxylation
AA: asymmetric aminohydroxylation
PHAL: phthalazine
PYR: pyrimidine
DP-PHAL: diphenylphthalazine
DPP: diphenylpyrazinopyridazine
AQN: anthraquinone
IND: 9-*O*-indolinylcarbamoyl
NMO: *N*-methylmorpholine *N*-oxide
NMM: *N*-methylmorpholine
LDH: layered double hydroxides
ABS-MC: poly(acrylonitrile-*co*-butadiene-*co*-styrene) microencapsulated
PEM-MC: phenoxyethoxymethyl-polystyrene microencapsulated

DHQ: dihydroquinine
DHQD: dihydroquinidine
DMAP: 4-(dimethylamino)pyridine
bmim: 1-butyl-3-methylimidazolium
emim: 1-ethyl-3-methylimidazolium
PEG: poly(ethylene glycol)
ee: enantiomeric excess

1
New Generations of Chiral Ligands

1.1
Substrate Scope and Limitations

Many years of thorough research and extensive screening of a large number of
compounds (>400) as chiral ligands have broadened the substrate scope of the
asymmetric dihydroxylation (AD) reaction and today only a handful of different ligands (Fig. 1) are needed to obtain good enantioselectivities in the AD reactions of a large variety of alkenes. A survey of the recommended ligands for
various classes of alkenes is given in Table 1 [1]. After the discovery of the extremely effective bis-Cinchona ligands with phthalazine (PHAL) or pyrimidine
(PYR) cores, the AD reaction generally gave good to excellent enantioselectivity for a wide range of alkenes [2]. The PHAL ligands worked well for alkenes
with aromatic substituents and *trans*-1,2-disubstituted alkenes. While the PYR
ligands were well suited for monosubstituted alkenes with aliphatic, sterically demanding substituents. Both, the PHAL and PYR ligands gave useful results
for some tetrasubstituted alkenes. The introduction of ligands having diphenylphthalazine (DP-PHAL), diphenylpyrazinopyridazine (DPP), or anthraquinone (AQN) cores further improved the substrate scope of the AD reaction [3,
4]. The DP-PHAL and DPP ligands have even broader scopes than the PHAL ligands. They are the optimal ligands for alkenes with aromatic substituents. The
AQN ligands are the ligands of choice for most alkenes with aliphatic substituents, especially for allylically substituted terminal alkenes. However, for alkenes
with aliphatic, sterically demanding substituents the PYR ligands have not been
superseded by any of the new ligands. *cis*-Alkenes are best served by the AQN,
DPP, or 9-*O*-indolinylcarbamoyl (IND) ligands, but remain challenging substrates for the AD reaction. A comparison of these ligands in the AD of selected
alkenes is given in Table 2 [2–4].

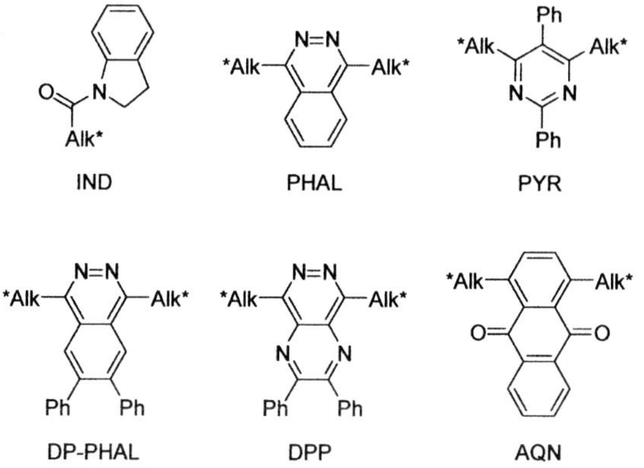

Fig. 1 Different cores of Cinchona alkaloid derived ligands used in the AD process (Alk*=DHQD or DHQ)

Table 1 Recommended ligands for the AD of different classes of alkenes

Alkene class	Substituent	Ligand	Enantiomeric excess range (%)
Monosubstituted	Aromatic	DPP, PHAL	70–97
	Aliphatic	AQN	
	Branched	PYR	
1,1-Disubstituted	Aromatic	DPP, PHAL	70–97
	Aliphatic	AQN	
	Branched	PYR	
cis-Disubstituted	Acyclic	IND	20–80
	Cyclic	DPP, AQN, PYR	
trans-Disubstituted	Aromatic	DPP, PHAL	90–99.8
	Aliphatic	AQN	
Trisubstituted		PHAL, DPP, AQN	90–99
Tetrasubstituted		PHAL, PYR	20–97

Table 2 Comparison of the enantiomeric excesses obtained with DHQD ligands in the AD of selected alkenes[a]

Alkene	PHAL	DPP	AQN	PYR	IND	Diol conf.
$H_{17}C_8$ ⌇	84	89	**92**	89		(R)
I ⌇	63	68	**83**	70		(S)
tBu ⌇	64	59		**92**		(R)
Ph ⌇	97	**99**	89	80		(R)
$H_{11}C_5$ ⌇	78	78	**85**	76		(R)
Ph ⌇	94	**96**	82	69		(R)
H_9C_4 ⌇ C_4H_9	97	96	**98**	88		(R,R)
Ph ⌇	97 (S,S)[b]	**98**	92			(R,R)
Ph ⌇ Ph	**99.8**					(R,R)
$H_{11}C_5$ ⌇ CO_2Et	**99**		**99**			(2S,3R)
Ph ⌇	35	68	45		**72**	(1R,2S)
(indene)	42	20	**63**	35		(1R,2S)
HO ⌇ OBz	64	**82**			31	(1S,2R)
⌇	98	98	**99**[c]	87		(R)
Ph ⌇	39			47		(R)

[a] The highest ee obtained in each case is in bold typeface
[b] Obtained from the DHQ ligand
[c] The same ees were obtained with the DP-PHAL ligand

1.2
Chiral Ligands for an Enantioselective Second Cycle

The discovery of the two catalytic dihydroxylation cycles present under homogeneous conditions was an important step in the development of today's highly successful AD process. While the first cycle is highly enantioselective, the second cycle proceeds with poor face selectivity, since it does not involve the chi-

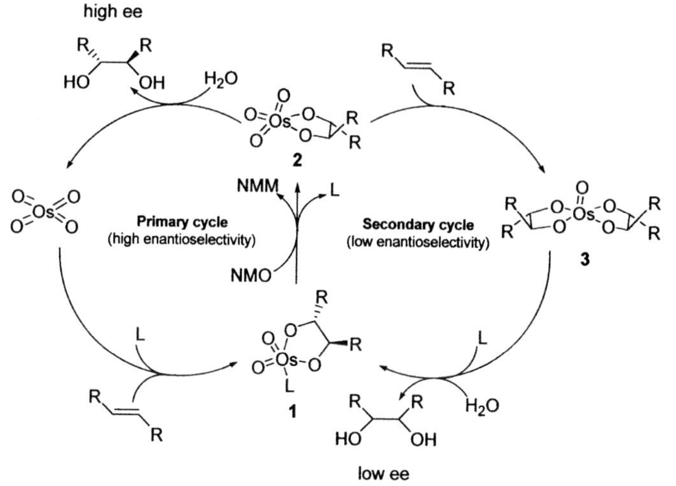

Scheme 1

ral ligand (Scheme 1). This obstacle was overcome (i) by slow addition of the alkene, which favors hydrolysis of the osmium(VIII) trioxoglycolate complex **2** over addition of a second alkene molecule, (ii) by addition of methanesulfonamide, which accelerates hydrolysis of **2**, and finally (iii) by introduction of the biphasic potassium ferricyanide cooxidant system.

Lately, new strategies capitalizing on the earlier unwanted secondary cycle have emerged. Sharpless, Fokin, and coworkers [5] have taken the first steps towards an AD proceeding through the secondary cycle. They discovered that certain classes of alkenes, Baylis–Hillman alkenes [6], and α,β-unsaturated amides [7, 8], and carboxylates [9], exhibited excellent reactivity in racemic aminohydroxylation independently of whether or not a ligand was added. Among these alkenes, unsaturated carboxylates constitute an extreme case. Racemic products were obtained even when a large excess of the chiral ligand was used suggesting that the reaction proceeded through the second catalytic cycle. The increased reactivity was explained by the presence of the carboxylate group, which facilitates the rate-limiting step, that is, the hydrolysis of the osmium(VI) bis(glycolate) **3**. On the basis of these findings, a new set of ligands for an enantioselective second cycle was designed. Vicinal hydroxysulfonamides have high binding constants for osmium thereby favoring the reaction to proceed through the second catalytic cycle. Furthermore, a free carboxylate group was important for a successful ligand, probably to ensure hydrolysis of **3**. The best ligands found so far are based on phenylisoserine **4** or threonine **5** (Table 3). Dihydroxylation of styrene with ligand **4** gave styrene diol in excellent yield (>99%) and with 42% ee. Methyl cinnamates were converted to the corresponding diols with ee values ranging from 48 to 70%.

Moreover, the secondary cycle was utilized to develop a new procedure for racemic osmium-catalyzed dihydroxylation. Dihydroxylation with citric acid as

Table 3 Asymmetric dihydroxylation of styrene and methyl cinnamates proceeding in the second catalytic cycle[a]

X	R	ee (%) (ligand 4), diol conf.	ee (%) (ligand 5), diol conf.	Ligand (mol%)
H	H	42, (R)		2
H	H	42, (R)	25, (S)	5
H	CO₂Me	53	48	2
NO₂	CO₂Me	40	70 (2R,3S)	2

[a] Performed in $tBuOH/H_2O$ (1:1) with 1.1 equivalents of NMO and 0.2 mol% of OsO_4

an additive in acidic media (pH 4–6) was successfully applied to the dihydroxylation of a range of electron-deficient alkenes [10].

2
Improved Reaction Conditions

During their efforts to employ molecular oxygen as the terminal oxidant in the AD process, the Beller group recognized that pH values in the range 11.2 –12.0 were essential to enhance hydrolysis of the osmium complex 3 formed from internal alkenes. Measurements of the pH during the dihydroxylation of *trans*-5-decene under standard AD conditions revealed a starting pH value of 12.2, which, in agreement with the overall stoichiometry (Scheme 2), continuously dropped during the reaction to reach a final value of 9.9. As suspected, the rate of product formation decreased with decreasing pH values. An experiment in which the pH was controlled by automatic titration (pH 12.0) led to a significant rate enhancement. Full conversion of the alkene was reached after less than 2 h compared to 34 h by applying standard AD conditions (without methansulfonamide). However, the enantioselectivity slightly decreased under the strong basic conditions compared to normal AD conditions. This decrease was explained by competition of hydroxide ions with the chiral ligand and consistent with this assumption, it could be overcome by higher ligand concentrations.

The rate enhancement due to pH control was most significant with tetrasubstituted alkenes (Table 4), making it possible to run the dihydroxylation of 2-methyl-3-phenyl-2-butene even at 0°C, which led to an increased enantioselectivity compared to earlier results (see Tables 2 and 4). However, in the case of α-methylstilbene, the constantly high pH favored the formation of higher oxidation products [11].

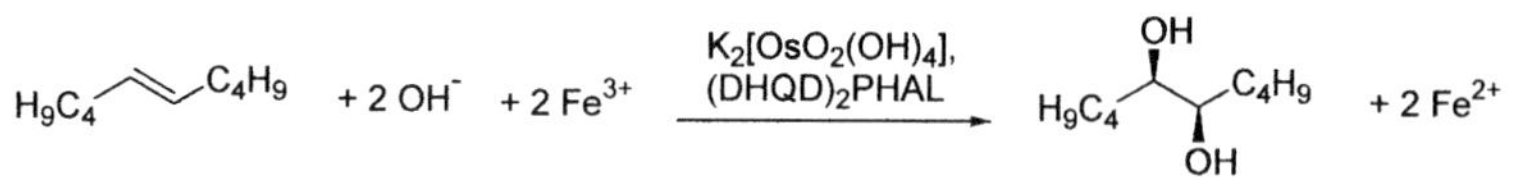

$$H_9C_4 \diagup \diagdown C_4H_9 + 2\ OH^- + 2\ Fe^{3+} \xrightarrow[\text{(DHQD)}_2\text{PHAL}]{K_2[OsO_2(OH)_4],} H_9C_4 \diagup \overset{\overset{\text{OH}}{|}}{\underset{\underset{\text{OH}}{|}}{}} \diagdown C_4H_9 + 2\ Fe^{2+}$$

Scheme 2

Table 4 The effect of the pH on the AD process

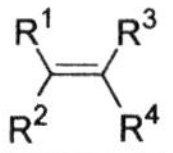

$$R^1 \diagdown \diagup R^3$$
$$R^2 \diagup \diagdown R^4$$

R^1	R^2	R^3	R^4	pH control	T (°C), t (h)	Ligand	Yield (%), ee (%)
C_4H_9	H	H	C_4H_9	–	25, 34	PHAL	94, 93
				–	25, 3[a]	PHAL	95, 93
				12.0	25, 1.8	PHAL	95, 90
				12.0	25, 1.7	PHAL[b]	96, 93
Ph	Me	H	Ph	–	25, 21	PHAL	82, 99[c]
				12.0	25, 1.5	PHAL	62, 99[c]
Ph	Me	Me	Me	–	25, 24	PHAL	28, 33
				12.0	25, 24	PHAL	95, 23
				12.0	0, 24	PHAL	71, 52
				12.0	0, 24	PYR	73, 61

3
New Cooxidants

As a consequence of the development of the *N*-methylmorpholine *N*-oxide (NMO) and later the potassium ferricyanide cooxidant systems the amounts of osmium tetroxide and chiral ligand used in the reaction could be considerably reduced. However, the method remains problematic for large-scale applications. The cooxidants for Os(VI) are expensive and large amounts of waste are produced (Table 5). Lately, several groups have addressed this problem and new re-oxidation processes for osmium(VI) species have been developed.

Recently, Bäckvall and coworkers described an improved H_2O_2 cooxidant system. They used H_2O_2 as terminal oxidant and *N*-methylmorpholine (NMM) and a flavin analogue as electron-transfer mediators [12, 13] (Scheme 3) to circumvent the overoxidation described earlier with H_2O_2 as the only cooxidant [14]. Flavin hydrogenperoxide selectively oxidized NMM to NMO, which in turn reoxidized osmium(VI) to osmium(VIII). The addition of tetraethylammonium acetate had a beneficial effect on the yield, probably due to enhanced hydrolysis of the osmium(VIII) trioxoglycolate **2**. This triple catalytic system could also be applied to the asymmetric dihydroxylation with the (DHQD)$_2$PHAL ligand.

Table 5 Comparison of several cooxidant systems in the asymmetric dihydroxylation ((DHQD)$_2$PHAL) of α-methylstyrene

Oxidants	Yield (%), ee (%)	Reaction conditions	TON	Waste (oxidant) (kg kg^{-1} diol)
K$_3$[Fe(CN)$_6$]	90, 94	0°C, K$_2$[OsO$_2$(OH)$_4$], tBuOH/H$_2$O	450	8.1
NMO	90, 33[a]	0°C, OsO$_4$, acetone/H$_2$O	225	0.88
NMM/flavin/H$_2$O$_2$	88, 99	rt, OsO$_4$, acetone/H$_2$O, Et$_4$NAc	44	0.25
(DHQD)$_2$PHAL/ flavin/H$_2$O$_2$	81, 90	rt, OsO$_4$, acetone/H$_2$O, Et$_4$NAc	41	0.04
PhSeCH$_2$Ph/O$_2$	89, 96	12°C, K$_2$[OsO$_2$(OH)$_4$], tBuOH/H$_2$O	222	0.16
PhSeCH$_2$Ph/air	87, 93	12°C, K$_2$[OsO$_2$(OH)$_4$], tBuOH/H$_2$O	48	0.16
O$_2$ (1 bar)	96, 80	50°C, K$_2$[OsO$_2$(OH)$_4$], tBuOH/aq. buffer	192	0
Air (20 bar)	95, 83	50°C, K$_2$[OsO$_2$(OH)$_4$], tBuOH/aq. buffer	950	0

[a] pClBz-DHQD
TON turnover number
rt room temperature

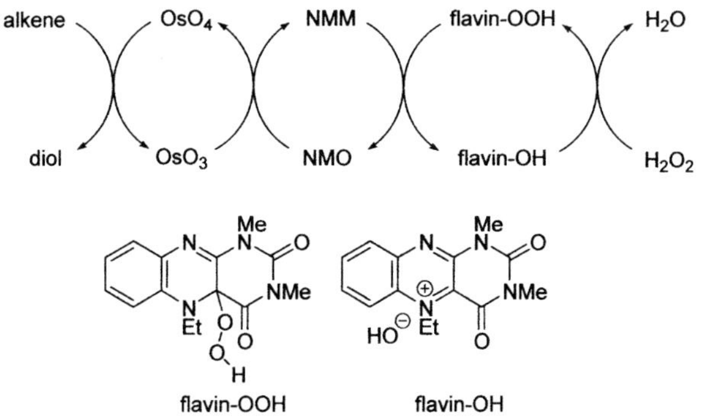

Scheme 3

However, since the catalytic system is homogenous, carefully adjusted reaction conditions were needed to circumvent the second nonselective catalytic cycle. Slow addition of the alkene and the hydrogen peroxide was necessary to obtain good enantioselectivities (Table 6) [13]. Recently, Bäckvall's group reported that the Cinchona alkaloid ligand participated in the reoxidation process and took the role of NMO in the catalytic cycle [15]. Versions of the triple catalytic system with vanadyl acetylacetonate replacing the flavin analogue [16] or *m*-CPBA as the terminal oxidant [17] have been developed and successfully applied to racemic dihydroxylation reactions.

Table 6 Comparison of the asymmetric dihydroxylation (($DHQD)_2PHAL$) with different cooxidant systems

	Alkenes			
	Ph⟋⟍	Ph⟋⟍Ph	Ph⟍⟋	⬡–Ph
Oxidants	Yield (%), ee (%)			
$K_3Fe(CN)_6$	>80, 97	>80, >99.5	90, 94	>80, 99
NMO	>80, 62	89, 88	90, 33	
NMM/flavin/H_2O_2	80, 95	94, 90	88, 99	50, 92
$(DHQD)_2PHAL/$ flavin/H_2O_2	75, 95	89, 90	81, 90	58, 70
$PhSe(O)CH_2Ph$	93, 97	84, 99	96, 97	95, 99
$PhSeCH_2Ph/O_2$			93, 97	
$PhSeCH_2Ph$/air			87, 93	
O_2 (1 bar)	49, 89	25, 90	96, 80	82, 90
Air (20 bar)	76, 87	89, 98	95, 83	88, 89

From economical and environmental points of view, air or dioxygen are the most attractive cooxidants. Several groups have addressed this problem without success [18–23], but in recent years Krief and Colaux–Castillo and the Beller group have independently developed cooxidant systems with dioxygen or air as the terminal oxidants. While Krief and Colaux–Castillo used selenoxides as cocatalysts, the Beller group was able to achieve oxidation of osmium(VI) to osmium(VIII) directly by dioxygen or air without the addition of cocatalysts.

Krief and Colaux–Castillo found that aryl selenoxides were efficient cooxidants in the AD process [24, 25]. The reaction is run under conditions similar to the AD-mix method with an aryl selenoxide (1.1 equivalents) in place of the ferricyanide (3.0 equivalents). Additionally, the amount of the K_2CO_3 could be reduced from 3.0 equivalents in the AD mix to 0.3 equivalents when selenoxides were applied as terminal oxidant. Diaryl selenoxides and selenoxides with electron-withdrawing aryl substituents were the best cooxidants. The choice of selenoxide had a strong influence on the reaction rate, but did only slightly affect the enantioselectivity (Table 7). Selenoxides are particularly attractive as cooxidants in the AD process, since the resulting selenides can be reoxidized by singlet oxygen, thus opening the possibility to use dioxygen as terminal oxidant. Krief's group was able to develop a cooxidant system in which substoichiometric amounts of a selenoxide were used as cocatalyst and dioxygen or air as terminal oxidant [25, 26]. The dioxygen originally in the triplet state had to be transferred to the active singlet state by visible light in the presence of a trace amount of Rose Bengal.

The Beller group was able to optimize the reaction conditions of the osmium-catalyzed dihydroxylation such that dioxygen or air could be used as cooxidants

Table 7 Asymmetric dihydroxylation of α-methyl styrene with different selenoxides as cooxidants

Selenoxide	Half-time (min)	ee (%)
PhCH$_2$Se(O)Me	810	94
p-MeO-PhSe(O)Me	163	90
PhSe(O)Me	163	88
PhCH$_2$Se(O)Ph	70	91
PhSe(O)Ph	56	90
p-Cl-PhSe(O)Me	56	90
p-NO$_2$-PhSe(O)Me	16	91

without additional cocatalysts [27–29]. To prevent overoxidation of the diol, a biphasic solvent system consisting of aqueous phosphate buffer and *tert*-butanol was used. Stereoselective dihydroxylation was achieved by the use of chiral dihydroquinine (DHQ)- or dihydroquinidine (DHQD)-based ligands. The observed enantioselectivities were somewhat lower compared to dihydroxylation with ferricyanide mainly due to the higher reaction temperature (50°C compared to 0°C). However, enantioselectivities close to the ceiling ee at 50°C were achieved when increased ligand concentrations were used to overcome competition between hydroxide ions and the chiral ligand. In contrast to the ligand-accelerated catalysis observed for dihydroxylations with other cooxidant systems, Beller and coworkers did observe a negative effect of ligand addition on the reaction rate. The reaction rate and the chemoselectivity were strongly influenced by the pH with an optimum pH range of 10.4–12.0 depending on the alkene structure. The catalyst activity was improved by increased dioxygen or air pressure without any significant decrease in chemoselectivity, which is especially important for the otherwise very slow dihydroxylations with air as cooxidant (Table 8). A number of different classes of alkenes (e.g., 1,1-disubstituted alkenes, 1,2-disubstituted alkenes, terminal aliphatic alkenes, tri- and tetrasubstituted alkenes, and functionalized alkenes) were shown to undergo dihydroxylation with good chemoselectivities. Alkenes like stilbene, cinnamic acid, and acrylic acid are still challenging substrates due to oxidative cleavage of the corresponding 1,2-diols.

4
Immobilization of Osmium

Although the AD process has found widespread use on the lab scale, industrial applications are obstructed owing to the toxicity and the high cost of the osmium catalyst and the risk of contamination of the products by toxic osmium residues. To address this issue, several research groups have developed immobilized osmium catalysts for use in the osmium-catalyzed dihydroxylations [30].

Table 8 Asymmetric dihydroxylation of α-methyl styrene with dioxygen or air as cooxidants

Cooxidant, pressure (bar)	TOF (h⁻¹)	ee (%)	Selec. (%)	Yield (%)	pH	[L], (mmol L⁻¹)	t (h)	L:Os	Ligand
O$_2$, 1	6	–	90	75	9.5	–	24	–	–
O$_2$, 1	15	–	92	92	10.4	–	12	–	–
O$_2$, 1	1	–	40	9	13.0	–	24	–	–
O$_2$, 1	12	75	96	96	10.4	1	16	1:1	(DHQD)$_2$PHAL
O$_2$, 1	10	80	96	96	10.4	3	20	3:1	(DHQD)$_2$PHAL
O$_2$, 1	–	88	98	98	10.4	100	24	1:1	(DHQD)$_2$PHAL
O$_2$, 10	120	76	93	71	10.4	0.8	24	15:1	(DHQD)$_2$PHAL
Air, 20	40	78	94	94	10.4	3	24	6:1	(DHQD)$_2$PHAL

TOF turnover frequency

Several heterogeneous osmium catalysts have been used in racemic dihydrox-ylations. In a series of patents, Michaelson and coworkers described the prepara-tion and the use of osmium catalysts heterogenized on supports such as MgO or Al$_2$O$_3$ [22, 31, 32]. The groups of Cainelli [33, 34] and Hermann [35] reported the use of amine-functionalized polymers as supports for osmium tetroxide. Their approach relies on the coordination of nitrogen to osmium, the same coordina-tion that is held responsible for the ligand acceleration effect. However, leaching tests by Jacobs showed that the filtrate contained active osmium [30]. According to the catalytic cycle (Scheme 1), the osmium is detached from the nitrogen dur-ing catalyst turnover, opening the possibility for osmium leaching from the pol-ymeric support. In an alternative approach, Jacobs et al. immobilized osmium tetroxide by addition to a silica-anchored tetrasubstituted alkene [36]. Leaching tests confirmed that the resulting osmium glycolate ester was stable towards hy-drolysis under the reaction conditions due to steric hindrance. The heterogene-ous osmium catalyst was also used in combination with electron-transfer me-diators such as titanocene dichloride and Na$_2$WO$_4$ immobilized on molecular sieves and layered double hydroxide (LDH), respectively, to allow the simulta-neous reoxidation of NMM to NMO by H$_2$O$_2$ [37]. The dihydroxylation reaction and the regeneration of NMO were performed in two physically separated ves-sels with discontinuous circulation of the liquid between the vessels.

The first heterogeneous osmium catalyst applicable for asymmetric dihy-droxylation reactions was described by Kobayashi and coworkers (Table 9, en-try 1) [38, 39]. Osmium tetroxide was enveloped in a polymer capsule by mi-croencapsulation techniques [40, 41]. The asymmetric dihydroxylation of *trans*-methylstyrene with poly(acrylonitrile-*co*-butadiene-*co*-styrene) microencap-sulated (ABS-MC) osmium tetroxide as catalyst, NMO as the cooxidant, and (DHQD)$_2$PHAL as the chiral ligand completed in 88% yield with 94% ee [38]. The catalyst and the chiral ligand were reused in five consecutive runs with-out loss of activity. However, the use of NMO as cooxidant required the slow

Table 9 Comparison of the asymmetric dihydroxylation ((DHQD)$_2$PHAL) with different heterogeneous osmium catalysts

		Alkenes			
		Ph⌒⟋	Ph⌒Ph	Ph⤙⟋	⬡-Ph
Entry	Catalyst (mol%), cooxidant	Yield (%), ee (%)			
1	ABS-MC OsO$_4$ (5), NMO	–	–	36, 85	64, 86
2	PEM-MC OsO$_4$ (5), K$_3$Fe(CN)$_6$	85, 78	66, >99	85, 76	85, 95
3	LDH-OsO$_4$ (1), NMO	94, 95	96, 99	89, 90	–
4	Resin-OsO$_4$ (1), NMO	92, 95	92, 99	93, 91	90, 91
5	Resin-OsO$_4$ (1), K$_3$Fe(CN)$_6$	89, 97	95, 99	92, 93	88, 99
6	Resin-OsO$_4$ (1), O$_2$	50, 89	20, 95	99, 84	80, 90
7	XAD-4-OsO$_4$ (1), K$_3$Fe(CN)$_6$	92, 95	94, >99	96, 89	93, 97
8	K$_2$OsO$_2$(OH)$_4$ (0.2), K$_3$Fe(CN)$_6$[a]	>80, 97	>80, >99.5	90, 94	>80, 99

[a] Sharpless KB, Amberg W, Bennani YL, Crispino GA, Hartung J, Jeong K-S, Kwong H-L, Morikawa K, Wang Z-M, Xu D, Zhang X-L (1992) J Org Chem 57:2768

addition of alkene. This problem was solved by the introduction of phenoxyethoxymethyl-polystyrene microencapsulated (PEM-MC) osmium tetroxide [39]. The PEM-MC osmium tetroxide could be applied in asymmetric dihydroxylations with the potassium ferricyanide cooxidant system in H$_2$O/acetone solutions. Microencapsulated osmium catalysts mostly gave lower yields and enantioselectivities than homogeneous reactions (Table 9, compare entries 2 and 8). Recently, Ley and coworkers showed that osmium tetroxide microencapsulated in a polyurea matrix could be used in non-asymmetric dihydroxylations with NMO as cooxidant, too [42].

Recently, Choudary and coworkers reported very efficient heterogeneous osmium catalysts immobilized by ion-exchange techniques [43, 44]. OsO$_4^{2-}$ was exchanged onto LDH and quaternary bounded ammonium groups anchored on organic resin or modified silica (Scheme 4). These osmium-exchanger catalysts displayed higher reactivity than K$_2$OsO$_2$(OH)$_4$ in dihydroxylation reactions and gave yields and enantioselectivities comparable with homogeneous dihydroxylations (Table 9, compare entries 3 and 8). When NMO was used as cooxidant, the LDH–osmium tetroxide could be reused three times without loss of activity. However, with potassium ferricyanide or dioxygen as cooxidant, rapid deactivation of the catalyst due to leaching of the OsO$_4^{2-}$ was observed. In contrast, res-

Scheme 4

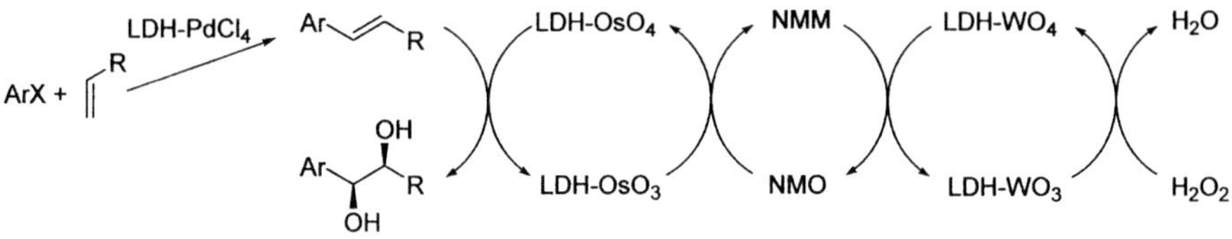

Scheme 5

in- and silica-bound osmium tetroxide performed with consistent results for a number of recycles regardless of the cooxidant (Table 9, entries 4–6).

In an approach conceptually related to that of Jacobs et al. (see above), Song and coworkers used macroporous, nonionic resins bearing residual vinyl groups (Amberlite XAD-4 or XAD-7) to immobilize osmium tetroxide [45]. The supported catalyst was used in ADs under standard $K_3Fe(CN)_6$ conditions and gave results comparable with homogeneous reactions (Table 9, compare entries 7 and 8). As a result of leaching of osmium, the catalytic activity decreased after three recycles causing increased reaction time. Unfortunately, the catalytic activity of the filtrate after recovery of the supported catalyst was not reported, thereby not ruling out a homogeneous reaction.

Choudary's group developed heterogeneous osmium catalysts for use in multifunctional catalysts for tandem Heck/AD, AD/N-oxidation, and even Heck/AD/N-oxidation reactions (Scheme 5) [46]. The $PdCl_4^{2-}$ and OsO_4^{2-} ions or OsO_4^{2-} and WO_4^{2-} ions, respectively, were simultaneously exchanged onto LDH to obtain the bifunctional catalysts, LDH–$(PdCl_4, OsO_4)$ and LDH–(OsO_4, WO_4). Simultaneous ion-exchange of all three anions onto a single LDH matrix gave the trifunctional catalyst, LDH–$(PdCl_4, OsO_4, WO_4)$. Tandem Heck coupling/AD reactions to obtain chiral 1,2-disubstituted diols from aryl iodides and terminal alkenes were catalyzed by the palladium and osmium sites of the trifunctional catalyst. Additionally, the tungsten sites of the catalyst acted as electron-transfer mediators for the selective reoxidation of NMM to NMO by H_2O_2. Under the influence of $(DHQD)_2PHAL$, chiral diols were achieved in good yields (>85%) with high enantioselectivities (>95%).

Recently, Yao showed that osmium tetroxide could be immobilized in an ionic liquid. The recyclability of osmium tetroxide was improved by the addition of DMAP. Both the catalyst and the ionic liquid were reused in six consecutive runs without significant reduction in yield [47]. Dihydroxylations in a solvent mixture of 1-butyl-3-methylimidazolium hexafluorophosphate ([bmim]PF_6), tert-butanol, and water with OsO_4 (2 mol%), DMAP (2.4 mol%), and NMO (1.1 equivalents) as cooxidant afforded diols in good yield (73–99% depending

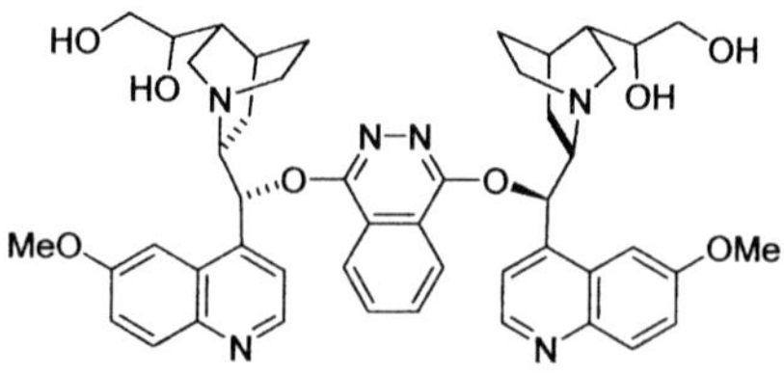

6

Fig. 2

on the substrate) after 40 h. Almost simultaneously, Yanada and Takemoto described the immobilization of osmium tetroxide in the ionic liquid 1-ethyl-3-methylimidazolium tetrafluoroborate ([emim]BF$_4$) without any cosolvent. Under these conditions, no significant reduction in yield (93%) was observed after five consecutive runs [48]. Shortly after the first reports on non-asymmetric osmium-catalyzed dihydroxylations in ionic liquids, two reports on asymmetric protocols appeared [49, 50]. Branco and Afonso reported the first application of K$_2$OsO$_2$(OH)$_4$ as catalyst and potassium ferricyanide as cooxidant in ionic liquid solvent systems [49]. Dihydroxylations were carried out in a biphasic [bmim]PF$_4$/water or a monophasic [bmim]PF$_4$/*tert*-butanol/water system under the influence of (DHQD)$_2$PHAL or (DHQD)$_2$PYR and compared to experiments performed in *tert*-butanol/water under similar conditions. In general, at least one ionic liquid solvent system resembled the yield and enantioselectivity obtained in *tert*-butanol/water. The recovered osmium catalyst in the ionic liquid was reused in nine successive runs before the yield and enantioselectivity slowly degraded. The osmium leaching from the ionic liquid to the aqueous and ethereal (product) phases was thoroughly investigated by inductively coupled plasma spectroscopy (ICP) showing that the aqueous phase generally contained 3–4% osmium (up to 14% after the first run) while the amount of osmium present in the organic phase was in the range of the detection limit (<3%). Song and coworkers applied ionic liquids [bmim]PF$_4$ and SbF$_6$ in acetone/water to the AD reaction with NMO as cooxidant [50]. The results were comparable with those obtained without ionic liquid. However, sever leaching of the osmium and the chiral ligand (DHQD)$_2$PHAL was observed during extraction with ether. Use of the tetraol **6** (Fig. 2) as chiral ligand improved the recyclability of both components and the yield started to decrease after three consecutive runs.

5
Immobilization of the Chiral Ligand

To reduce the cost of the AD process immobilization of the chiral Cinchona alkaloid-derived ligands has attracted attention, too. Several approaches to address this problem have been reported [51]. The chiral ligand has been attached to solid supports comprising organic polymers or modified silica. After completion of the AD reaction under heterogeneous conditions, the ligand was isolated by filtration. Alternatively, the ligand was immobilized by attachment to a polymeric unit, which was soluble in the reaction medium allowing AD under homogeneous conditions. The ligand was separated by filtration after precipitation of the ligand by addition of a solvent.

Unfortunately, the immobilization of the alkaloid ligands did not result in the simultaneous immobilization of the osmium catalyst due to the weak binding of cinchonidine to the osmium complexes. In most studies, leaching of the osmium catalyst was reported and supplementation of the osmium catalyst was necessary after recovery of the immobilized chiral ligand. In other studies, the recycled ligands were used without further addition of the osmium catalyst resulting in reduced yields or longer reaction times.

5.1
Insoluble Organic Polymers

The first report on heterogenized chiral ligands for use in AD reactions came in 1990 from Sharpless's laboratory [52]. The alkaloid ligands were incorporated into the polymer backbone with anchoring points either at the O-9 substituent (Fig. 3, polymer 7) or at the quinuclidine moiety (Fig. 3, polymer 8). Additionally, the alkaloid copolymers had long spacer groups to minimize the steric influences of the polymeric chain on the alkaloid ligand. The polymer-bound ligands 7 and 8 were used in the AD of trans-stilbene with good enantioselectivities and reasonable reaction rates (Table 10).

Lohray and coworkers prepared a polymer-bound ligand conceptually similar to polymer 7 (Fig. 3, polymer 9a) [53]. Polymers with 10% of DHQD anchored on the polystyrene backbone were the most efficient catalysts. In contrast to these observations, Song and coworkers reported good yield and high ee with the homopolymer 9b having a 100% DHQD loading (Table 10) [54].

Almost simultaneously to the publication by Kim and Sharpless, Salvadori and coworkers reported the use of polymers 10a–c related to 7, in which the quinuclidine moiety of the alkaloid ligand was attached to the polymer without separation by a spacer [55]. Studies of the influence of the nature of the polymeric ligand on the AD process showed that a low alkaloid content (below 15 mol%) was essential to obtain high yields of the diol. However, in the AD of trans-stilbene these ligands gave considerably lower enantioselectivities but better reactions rates compared to polymer 8 (Table 10). A new crosslinked polystyrene-bound ligand 7 with a spacer group gave improved enantioselectivity under NMO conditions [56]. In contrast, this polymeric ligand was inactive with $K_3Fe(CN)_6$ probably because the polymer collapsed in the protic polar solvent. This problem was overcome by changing the copolymers to hydroxyethyl methacrylate and ethylene glycol dimethacrylate (polymer 12a) [57]. AD reaction with a number of terminal and internal, aliphatic and aromatic alkenes under ferrocyanide conditions proceeded in many cases with enantioselectivities comparable to those achieved with soluble ligands. Moreover, polymeric ligands

Table 10 Comparison of immobilized ligands in the AD of *trans*-stilbene

	Ligand, mol%	Cooxidant	Yield (%)	ee (%)	T (°C)	t (h)	mol% Os
Insoluble polymers	7, 25	$K_3Fe(CN)_6$	91	86	rt	48	1.25
	8, 25	NMO	81–87	85–93	10	48–72	1
	9a, 13	NMO	82	85	0	24	1
	9b, 25	$K_3Fe(CN)_6$	80	89	10	20	1
	10c, 3	NMO	86	45	0	21	0.1
	11, 11	NMO	85	87	0	7	0.5
	12a, 25	$K_3Fe(CN)_6$	70	95	0	24	1.25
	12c, 10–25	$K_3Fe(CN)_6$	90	>99	0	20	0.5–1
Inorganic	13a, –	$K_3Fe(CN)_6$	96	80	20	24	–
	14a, 2	$K_3Fe(CN)_6$	88	>99	10	25	1
	14b, 1	$K_3Fe(CN)_6$	–	>99.5	0	24	0.5
	14b, 1	$K_3Fe(CN)_6$	92	99	rt	12–24	1
	14c, 1	NMO	96	99	rt	12–24	1
	14c, 1	$K_3Fe(CN)_6$	95	99	rt	12–24	1
	15a, 1	$K_3Fe(CN)_6$	97	>99	0	14	1
	16a, 2	$K_3Fe(CN)_6$	77	99	–	–	0.5–1
Soluble polymers	20, 25	NMO	88	89	4	5	
	21, 10	$K_3Fe(CN)_6$	95	99	–	–	0.5

rt room temperature

with altered *O*-9 substituents such as phenanthryl (polymer **12b**) and phthalazine (polymer **12c**) were investigated and found to give further improved ee
[58–62]. However, spectroscopic analysis of the washings after prolonged extraction (7 days) of the crude polymer **12c** showed significant absorption corresponding to the chiral monomer. Consequently, the AD reactions may, at least
in part, be catalyzed by the homogeneous monomer instead of the heterogeneous polymeric ligand. Similar problems seemed to occur with polymeric ligands
obtained by copolymerization of allylic alkaloidic monomers with electron-deficient monomers such as methyl methacrylate, hydroxyethyl methacrylate, and
ethylene glycol dimethacrylate [63–65]. On the basis of investigations of these
polymers after continuous extraction, the groups of Sherington [66] and Salvadori [62] suggested that the polymeric material did not consist of copolymers
but rather contain physically trapped alkaloidic monomer.

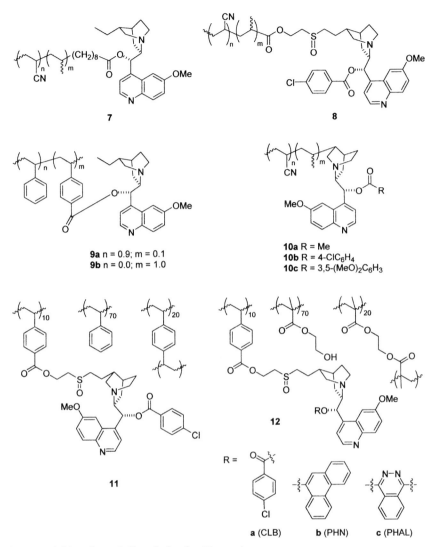

Fig. 3 Insoluble polymeric ligands for the AD reaction

5.2
Inorganic Supports

Lohray and coworkers reported the first application of silica gel-supported Cinchona alkaloids in AD in 1996 [67]. A 3,6-DHQ$_2$-pyridazine derivative was linked to a silica gel support with an attachment point at one of the quinuclidine moieties (Fig. 4, catalyst **13**). The alkaloidic ligand was expected to bind to the silica surface resulting in better availability of the active site compared to polymeric ligands. Enantioselectivities obtained with 13, however, were low compared to the best polymeric catalysts (Table 10). Shortly after, Song and coworkers described a more successful silica gel-supported DHQ$_2$PHAL ligand 14a with both quinuclidine moieties attached to the solid support [68]. In 2001, further improvements were reported almost simultaneously by the groups of Kim [69] and Crudden [70]. DHQ$_2$PHAL or DHQD$_2$PHAL ligands, respectively, were grafted onto mesoporous molecular sieves of the SBA-15 type leading to supported ligands 14b and 15a. In contrast to regular amorphous silica gel, SBA-15 has a uniform pore size and shape and a larger surface area. These special characteristics of SBA-15 were preserved after grafting of the alkaloid ligands. In general, ligands 14b and 15a gave similar enantioselectivities comparable to the results obtained in homogenous dihydroxylations for a range of alkenes. Results for the AD of *trans*-stilbene are shown in Table 10. Comparison of 15a with the corresponding silica gel-supported ligand 15b and homogenous DHQD$_2$PHAL at low catalyst loadings (0.1 mol%) showed that the SBA-supported ligand was slightly superior with respect to yield (54% compared to 46% and 44%, respectively) [70]. The group of Choudary used another type of mesoporous molecular sieve (MCM-41) in AD reactions with NMO, potassium ferricyanide, or dioxygen as cooxidants [71]. The MCM-41-supported ligand 14c gave results similar to the SBA-15-supported ligands (Table 10).

Recently, a bifunctional, silica gel-supported catalyst, containing an alkaloid ligand 14d and a palladium(0)complex, was applied to the dihydroxylation of disubstituted alkenes formed in situ from aryl iodides and terminal alkenes in a single-pot procedure [72].

Bolm and coworkers chose another strategy for attachment of the alkaloid ligand to the inorganic support. Instead of using an attachment point at the quinuclidine moiety, the 9-*O* substituent was connected to the support to minimize steric interactions between the support and the catalytically active site. The PYR, DPP [73], and AQN [74] cores of bisalkaloid ligands were modified and linked to silica gel (ligand 16a, 17a, and 18a, respectively). Enantioselectivities obtained with 16a and 17a compared well with the corresponding homogeneous systems, while 18a gave lower values.

5.3
Soluble Organic Polymers

In another approach, Bolm's group utilized soluble organic polymers to immobilize bisalkaloid ligands with PYR, DPP [75], and AQN [74] cores. MeO-poly(ethylene glycol) (MeO-PEG)-bound ligands 16b and 17b (Fig. 4) were soluble under ferricyanide reaction conditions and AD reactions (1 mol% OsO$_4$, 1 mol% ligand) gave good yields (84–92%) after short reaction times (<5 h) and enantioselectivities comparable with the corresponding non-polymer-bound ligands. After the reaction was completed, the ligands were precipitated by addition of *tert*-butyl methyl ether and isolated by filtration with more than 98% recovery. However, catalysis by reused ligands afforded slightly reduced enanti-

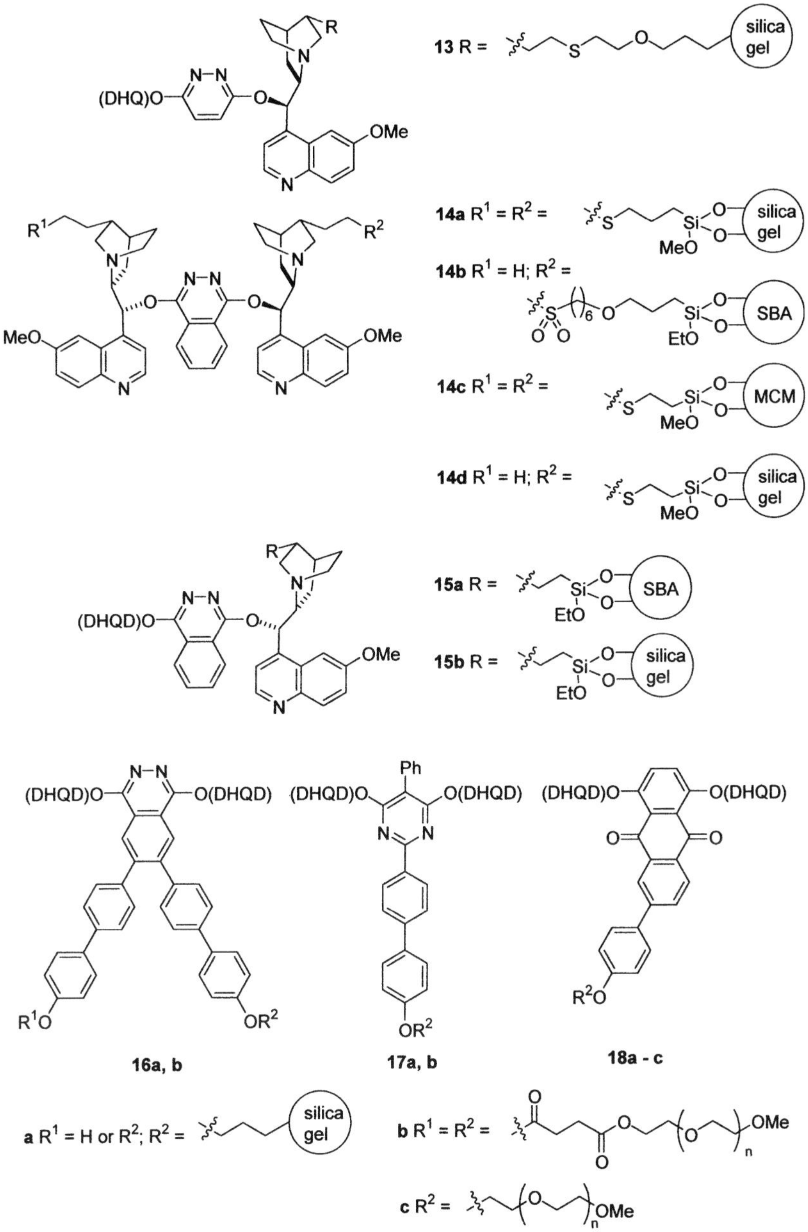

Fig. 4 Ligands immobilized on inorganic supports and soluble organic polymers

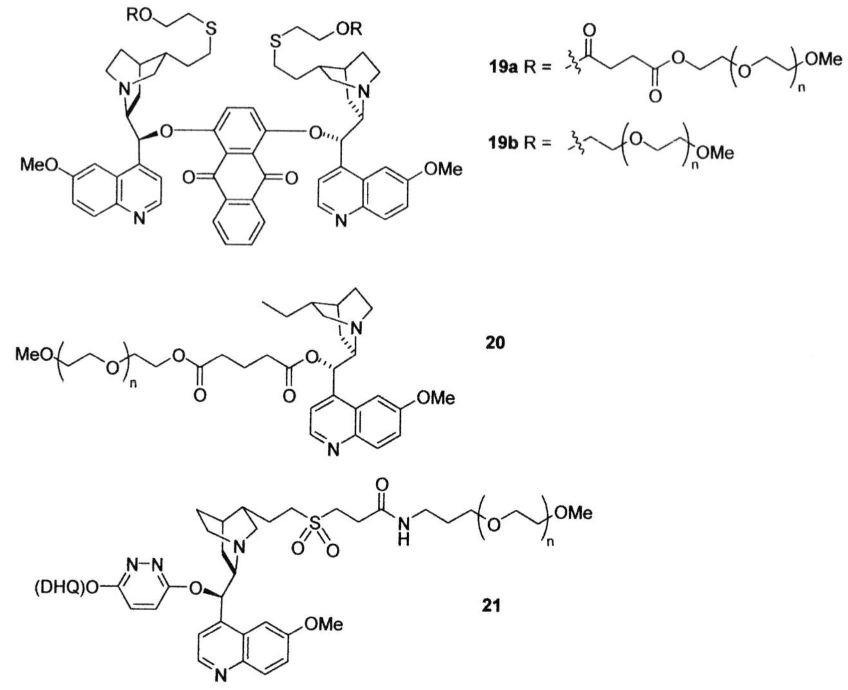

Fig. 5 Ligands immobilized on soluble organic polymers

oselectivities, probably due to loss of the alkaloid by hydrolysis of the ester attachment to the polymer [75]. Alkaloid ligands with AQN cores were bound to
MeO-PEG with ester and ether attachments, **18b** and **18c**, respectively (Fig. 4)
[74]. These ligands gave good enantioselectivities with respect to values reported for homogeneous reaction using immobilized ligands (80–84% ee, compared
to 83% ee) for the AD of allyl iodide but lower enantioselectivities for indene
(54–57% ee, compared to 63% ee). Interestingly, comparison of catalysts **18b**
and **18c** with the quinuclidine-attached analogues **19a** and **19b** (Fig. 5) showed
no significant influence of the attachment mode on the enantioselectivity. In
most cases, the quinuclidine-attached analogues, **19a** and **19b**, were slightly superior to the analogues linked at the spacers, **18b** and **18c**. At almost the same
time, Han and Janda reported on the use of the MeO-PEG-bound mono- and
bisalkaloid ligands **20** [76] and **21** [77] (Fig. 5). ADs of several 1,2-disubstitut-
ed alkenes using the bisalkaloid ligand **21** containing a PHAL core completed
with enantioselectivities comparable to their unsupported, homogeneous counterparts.

6
Conclusion

The AD has gained an enormous appreciation among synthetic chemists and the number of applications of the AD reaction is rapidly increasing. This review concentrates on the recent efforts made to further increase the applicability of the AD process with emphasis on large-scale application. In a future perspective, the combination of several of the concepts described above may lead to exciting improvements. For example, the development of second-cycle ligands may possibly allow the simultaneous immobilization of the chiral ligand and the osmium catalyst. Used in combination with dioxygen or H_2O_2 as reoxidants, industrial applications may be within reach.

References

1. Kolb HC, Sharpless KB (1998) Asymmetric dihydroxylation. In: Beller M, Bolm C (eds) Transition metals in organic synthesis. Wiley-VCH. Weinheim, New York, p 220
2. Kolb HC, VanNieuwenhenze MS, Sharpless KB (1994) Chem Rev 94:2483
3. Becker H, King B, Taniguchi M, Vanhessche KPM, Sharpless KB (1995) J Org Chem 60:3940
4. Becker H, Sharpless KB (1996) Angew Chem Int Ed 35:448
5. Andersson MA, Epple R, Fokin VV, Sharpless KB (2002) Angew Chem Int Ed 41:472
6. Pringle W, Sharpless KB (1999) Tetrahedron Lett 40:5151
7. Rubin AE, Sharpless KB (1997) Angew Chem Int Ed 36:2637
8. Gontcharov AV, Liu H, Sharpless KB (1999) Org Lett 1:783
9. Fokin VV, Sharpless KB (2001) Angew Chem Int Ed 40:3455
10. Dupau P, Epple R, Thomas AA, Fokin VV, Sharpless KB (2002) Adv Synth Catal 344:421
11. Mehltretter GM, Dobler C, Sundermeier U, Beller M (2000) Tetrahedron Lett 41:8083
12. Bergstad K, Jonsson SY, Bäckvall JE (1999) J Am Chem Soc 121:10424
13. Jonsson SY, Färnegårdh K, Bäckvall JE (2001) J Am Chem Soc 123:1365
14. Milas NA, Sussman S (1936) J Am Chem Soc 58:1302
15. Jonsson SY, Adolfsson H, Bäckvall JE (2001) Org Lett 3:3463
16. Éll AH, Jonsson SY, Börje A, Adolfsson H, Bäckvall JE (2001) Tetrahedron Lett 42:2569
17. Bergstad K, Piet JJN, Bäckvall JE (1999) J Org Chem 64:2545
18. Cairns JF, Roberts HL (1968) J Chem Soc C 640
19. Celanese Corp (1966) GB-B 1028940
20. Michaelson RC, Austin RG (Exxon Corp) (1982) EP0077201
21. Myers RS, Michaelson RC, Austin RG (Exxon Corp) (1984) US4496779
22. Michaelson RC, Austin RG (Exxon Corp) (1985) US4533772
23. Austin RG, Michaelson RC, Myers RS (1985) In: Augustine RL (ed) Catalysis of organic reactions. Dekker, New York, p 269
24. Krief A, Castillo-Colaux C (2001) Synlett 501
25. Krief A, Colaux-Castillo C (2002) Pure Appl Chem 74:107
26. Krief A, Colaux-Castillo C (1999) Tetrahedron Lett 40:4189
27. Döbler C, Mehltretter G, Beller M (1999) Angew Chem Int Ed 38:3026
28. Döbler C, Mehltretter GM, Sundermeier U, Beller M (2000) J Am Chem Soc 122:10289
29. Döbler C, Mehltretter GM, Sundermeier U, Beller M (2001) J Organomet Chem 621:70
30. Severeyns A, De Vos DE, Jacobs PA (2002) Top Catal 19:125
31. Michaelson RC, Austin RG, White DA (Exxon Corp) (1983) US4413151
32. Michaelson RC, Austin RG, White DA (Exxon Corp) (1984) US4486613
33. Cainelli G, Contento M, Manescalchi F, Plessi L (1989) Synthesis 45

34. Cainelli G, Contento M, Manescalchi F, Plessi L (1989) Synthesis 47
35. Herrmann WA, Kratzer RM, Blümel J, Friedrich HB, Fischer RW, Apperly DC, Mink J, Berkesi O (1997) J Mol Catal A 120:197
36. Severeyns A, De Vos DE, Fiermans L, Verpoort F, Grobet PJ, Jacobs PA (2001) Angew Chem Int Ed 40:586
37. Severeyns A, De Vos DE, Jacobs PA (2002) Green Chem 4:380–384
38. Kobayashi S, Endo M, Nagayama S (1999) J Am Chem Soc 121:11229
39. Kobayashi S, Ishida T, Akiyama R (2001) Org Lett 3:2649
40. Kobayashi S, Nagayama S (1998) J Am Chem Soc 120:2985
41. Kobayashi S, Endo M, Nagayama S (1998) J Org Chem 63:6094
42. Ley SV, Ramarao C, Lee A-L, Østergaard N, Smith SC, Shirley IM (2003) Org Lett 5:185
43. Choudary BM, Chowdari NS, Kantam ML, Raghavan KV (2001) J Am Chem Soc 123:9220
44. Choudary BM, Chowdari NS, Jyothi K, Kantam ML (2002) J Am Chem Soc 124:5341
45. Yang JW, Han H, Roh EJ, Lee S, Song CE (2002) Org Lett 4:4685
46. Choudary BM, Chowdari NS, Madhi S, Kantam ML (2001) Angew Chem Int Ed 40:4619
47. Yao Q (2002) Org Lett 4:2197
48. Yanada R, Takemoto Y (2002) Tetrahedron Lett 43:6849
49. Branco LC, Afonso CAM (2002) Chem Commun 3036
50. Song CE, Jung D, Roh EJ, Lee S, Chi DY (2002) Chem Commun 3038
51. Bolm C, Gerlach A (1998) Eur J Org Chem 21
52. Kim BM, Sharpless KB (1990) Tetrahedron Lett 31, 3003
53. Lohray BB, Thomas A, Chittari P, Ahuja JR, Dhal PK (1992) Tetrahedron Lett 33:5453
54. Song CE, Roh EJ, Lee S, Kim IO (1995) Tetrahedron Asymmetry 6:2687
55. Pini D, Petri A, Nardi A, Rosini C, Salvadori P (1991) Tetrahedron Lett 32:5175
56. Pini D, Petri A, Nardi A, Salvadori P (1993) Tetrahedron Asymmetry 4:2351
57. Pini D, Petri A, Nardi A, Salvadori P (1994) Tetrahedron 50:11321
58. Petri A, Pini D, Rapaccini S, Salvadori P (1995) Chirality 7:580
59. Petri A, Pini D, Salvadori P (1995) Tetrahedron Lett 36:1549
60. Salvadori P, Pini D, Petri A (1997) J Am Chem Soc 119:6929
61. Petri A, Pini D, Rapaccini S, Salvadori P (1999) Chirality 11:745
62. Salvadori P, Pini D, Petri A (1999) Synlett 1181
63. Lohray BB, Nandanan E, Bhushan V (1994) Tetrahedron Lett 35:6559
64. Song CE, Yang JW, Ha HJ, Lee S (1996) Tetrahedron Asymmetry 7:645
65. Nandanan E, Sudalai A, Ravindranathan T (1997) Tetrahedron Lett 38:2577
66. Canali L, Song CE, Sherrington DC (1998) Tetrahedron Asymmetry 9:1029
67. Lohray BB, Nandanan E, Bushan V (1996) Tetrahedron Asymmetry 7:2805
68. Song CE, Yang JW, Ha HJ (1997) Tetrahedron Asymmetry 8:841
69. Lee HM, Kim S-W, Hyeon T, Kim BM (2001) Tetrahedron Asymmetry 12:1537
70. Motorina I, Crudden CM (2001) Org Lett 3:2325
71. Choudary BM, Chowdari NS, Jyothi K, Kantam ML (2002) Catal Lett 82:99
72. Choudary BM, Chowdari NS, Jyothi K, Kumar NS, Kantam ML (2002) Chem Commun 586
73. Bolm C, Maischak A, Gerlach A (1997) Chem Commun 2353
74. Bolm C, Maischak A (2001) Synlett 93
75. Bolm C, Gerlach A (1997) Angew Chem Int Ed 36:741
76. Han H, Janda KD (1996) J Am Chem Soc 118:7632
77. Han H, Janda KD (1997) Tetrahedron Lett 38:1527

Chapter 20.2
Aminohydroxylation of Carbon–Carbon Double Bonds

Annette Bayer

Department of Chemistry, University of Tromsø,
9037, Tromsø, Norway
e-mail: annette.bayer@chem.uit.no

Keywords: Asymmetric aminohydroxylation, Osmium tetroxide, Cinchona alkaloid, Ligand-accelerated catalysis, Immobilization, Abbreviations

1	**Introduction**	44
2	**Catalytic Asymmetric Aminohydroxylation**	44
2.1	Nitrogen Source	46
2.2	Chiral Ligand and Solvent	47
2.3	Substrate	49
2.4	Supported Ligands	51
3	**On the Mechanism of Catalysis in the AA Process**	52
4	**Investigations of Chemoselectivity**	54
5	**Synthetic Applications of the Chiral Aminoalcohols**	56
5.1	Cinnamate, Arylacrylates, and Related Compounds	56
5.2	Styrenes and Related Compounds	61
5.3	Conjugated Esters	62
5.4	Miscellaneous	65
5.5	Diastereoselectivity in the AA Reaction	67
6	**Conclusion**	69
	References	69

Abbreviations

AD	asymmetric dihydroxylation
AA	asymmetric aminohydroxylation
Boc	*t*-butoxycarbonyl
Chloramine-T	*N*-chloro-*N*-sodio-*p*-toluenesulfonamide
Chloramine-M	*N*-chloro-*N*-sodio-methylsulfonamide
Cbz	benzyloxycarbonyl
PHAL	phthalazine
PYR	pyrimidine
AQN	anthraquinone

DHQ   dihydroquinine
DHQD   dihydroquinidine
Teoc   2-(trimethylsilyl)ethoxycarbonyl
ee   enantiomeric excess

1
Introduction

In 1975 Sharpless and coworkers discovered the stoichiometric aminohydrox-ylation of alkenes by alkylimido osmium compounds leading to protected vici-nal aminoalcohols [1, 2]. Shortly after, an improved procedure was reported em-ploying catalytic amounts of osmium tetroxide and a nitrogen source (N-chlo-ro-N-metallosulfonamides or carbamates) to generate the active imido osmium species in situ [3–8]. Stoichiometric enantioselective aminohydroxylations were first reported in 1994 [9]. Finally, in 1996 the first report on a catalytic asymmet-ric aminohydroxylation (AA) was published [10]. During recent years, several reviews have covered the AA reaction [11–16].

2
Catalytic Asymmetric Aminohydroxylation

The AA reaction is closely related to the asymmetric dihydroxylation (AD). Alkenes are enantioselectively converted to protected β-aminoalcohols (Scheme 1) by *syn*-addition of osmium salts under the influence of the chiral bis-Cinchona ligands known from the AD process (see Chap. 20.1). As for the AD reaction, a cooxidant is needed to regenerate the active osmium species. But in the AA process the cooxidant also functions as the nitrogen source. Since two different heteroatoms are transferred to the double bond, regioselectivity be-comes an important selectivity issue in addition to enantioselectivity. Moreo-ver, chemoselectivity has to be addressed due to the possible formation of the

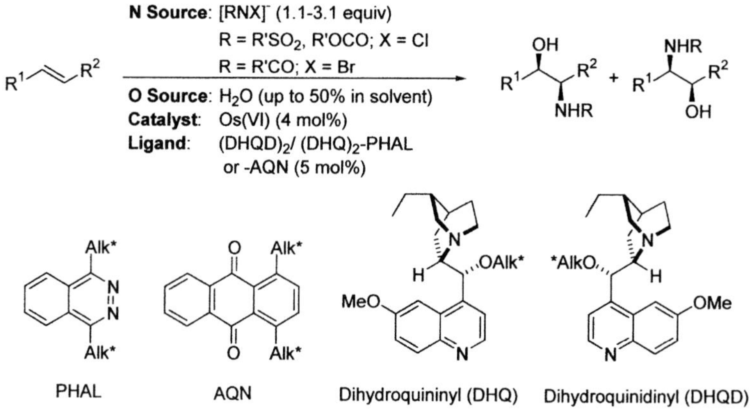

N Source: [RNX]⁻ (1.1-3.1 equiv)
R = R'SO₂, R'OCO; X = Cl
R = R'CO; X = Br
O Source: H₂O (up to 50% in solvent)
Catalyst: Os(VI) (4 mol%)
Ligand: (DHQD)₂/ (DHQ)₂-PHAL or -AQN (5 mol%)

PHAL   AQN   Dihydroquininyl (DHQ)   Dihydroquinidinyl (DHQD)

Scheme 1

diol as a byproduct. The outcome of the AA reaction with respect to yield, regioselectivity, and enantioselectivity strongly depends on several reaction parameters such as the source of nitrogen (Table 1), the solvent, the chiral ligand (Table 2), and the structure of the substrate.

Table 1 Comparison of various AA procedures[a]

Substrate	N source	Solvent, temp. (°C)	Yield (%), ee (%) of A	A:B	Time (h)
R^1=Ph;	TsNClNa	MeCN/H_2O, rt	64, 81	–	3
R^2=CO_2Me	MsNClNa	nPrOH/H_2O, rt	65, 95	10:1	3
	CbzNClNa	nPrOH/H_2O, rt	65, 94	–	0.6
	EtONClNa	nPrOH/H_2O, rt	78, 99	–	0.75
	BocNClNa	nPrOH/H_2O, 0-rt	79, 80	–	3
R^1=Ph;	TsNClNa	MeCN/H_2O, rt	–, 60	3.2:1	3.5
R^2=CO_2iPr	MsNClNa	MeCN/H_2O, rt	–, 89	9.0:1	0.7
	MsNClNa	nPrOH/H_2O, rt	65, 94	19:1	3.5
	TeocNClNa	nPrOH/H_2O, rt	70, 99	>49:1	0.5
	AcNBrLi	tBuOH/H_2O, 4	72, 99	>20:1	20
	PhCONBrLi	MeCN/H_2O, 0	43, 91	5.2:1	10
	TriazNClNa	EtOH/H_2O, rt	97, 99	>20:1	12
R^1=R^2=Ph	TsNClNa	tBuOH/H_2O, rt	64, 78	–	3
	MsNClNa	nPrOH/H_2O, rt	71, 75	–	16
	CbzNClNa	nPrOH/H_2O	92, 91	–	3.0
	AcNBrK	nPrOH/H_2O, 4	50, 94	–	–
	TriazNClNa	EtOH/H_2O, rt	86, 87	–	12
Cyclohexene	TsNClNa	MeCN/H_2O, rt	64, 45	–	6
	MsNClNa	nPrOH/H_2O, rt	49, 66	–	18
	CbzNClNa	nPrOH/H_2O, rt	51, 63	–	2.0
	TriazNClNa	EtOH/H_2O, rt	92, 79	–	12
R^1=Ph;	TsNClNa	–	low, 50–70	2:1	–
R^2=H	CbzNClNa	nPrOH/H_2O	60, 89	5:1	1.5
	BocNClNa	nPrOH/H_2O, 0	58, 94	4:1	1
	AcNBrLi	nBuOH/H_2O, 4	72, 91	1.1:1	–
	AcNBrLi	MeCN/H_2O, 4	55, 88	1:6.1	–
	TriazNClNa	EtOH/H_2O, rt	74, 86	5.0:1	12
R^1=H;	TsNClNa	MeCN/H_2O	–, 46	–	–
R^2=CO_2Et	CbzNClNa	nPrOH/H_2O, 25	89, 84	–	0.8
	AcNBrLi	tBuOH/H_2O, 4	46, 89	>20:1	–

[a] The highest ee achieved for each substrate is in bold
rt room temperature

2.1
Nitrogen Source

In the first reported procedure for the AA reaction, Chloramine-T (*N*-chloro-*N*-sodio-*p*-toluenesulfonamide) was used as cooxidant/nitrogen source in combination with the (DHQ)$_2$PHAL and (DHQD)$_2$PHAL ligands [10]. The reactions were performed in acetonitrile/water (1:1) or *tert*-butanol/water (1:1) using K$_2$OsO$_2$(OH)$_4$ (4 mol%) as osmium precursor, the chiral ligands (DHQ)$_2$PHAL or (DHQD)$_2$PHAL (5 mol%), and an excess of Chloramine-T (3 equivalents). Enantioselectivities in the range of 50–81% and 33–48% ee were obtained for *trans*- and *cis*-1,2-disubstituted alkenes, respectively. After the initial report, the introduction of other sulfonamides [17], carbamates [18–21], amides [22–24], and amino-substituted aromatic heterocycles [25] as nitrogen sources led to a rapid improvement in the scope and selectivity (Table 1). It was soon discovered that the methanesulfonamide-derived chloramine salt (Chloramine-M) was a superior nitrogen source. First of all, the chemo-, regio-, and enantioselectivities (63–95% ee) of the AA process were improved probably due to the smaller substituent at sulfur. Moreover, the *N*-mesyl products are more readily purified by recrystallization than the *N*-tosyl products obtained with Chloramine-T. Additionally, the residual methanesulfonamide was easily removed by extraction or sublimation [17]. In a further development of the AA process, benzyl (Cbz) [18], *tert*-butyl (Boc) [21, 20, 26], ethyl [18], and 2-(trimethylsilyl)ethyl (Teoc) carbamates [19] were successfully used as nitrogen sources leading to improved enantioselectivities and scope (aromatic and electron-deficient alkenes). Again, the small Teoc substituent on nitrogen led to higher reactivity than Cbz and Boc substituents and good regio- and enantioselectivities. In the case of TeocNNaCl as nitrogen source, the catalyst loading could be reduced to 2 mol% K$_2$OsO$_2$(OH)$_4$ and ligand loading to 2.5 mol% without loss in yield, regioselectivity, and enantioselectivity. A further advantage of the carbamate-AAs was the numerous methods for the removal of the Teoc (nucleophilic), Cbz (reductive), and Boc groups (acidic) under mild conditions. Whereas an excess of the cooxidant/nitrogen source is needed in the AA procedures described above, a nearly stoichiometric amount (1.1 equivalents) of alkali metal salts (Li$^+$ or K$^+$) of *N*-bromo carboxamides suffice for complete conversion [22, 23]. Additionally, the asymmetric induction was improved for many substrates when *N*-bromoacetamide was used as nitrogen source. Attempts to extend the scope of the nitrogen source to amino-substituted heterocycles led to the use of adenine derivatives in racemic aminohydroxylations [27, 28]. Later, an enantioselective procedure was achieved by using aminopyrimidines and aminotriazines as nitrogen source [29]. The use of intramolecular nitrogen sources has been exploited by Donohoe and coworkers as described later [30, 31].

  Some of the nitrogen sources mentioned above (e.g., Chloramine-T and *N*-bromoacetamide) are commercially available. In other cases, the chloramine salts have to be prepared in situ by treating the amide with *tert*-butyl hypochlorite [32] and sodium hydroxide [15]. It has been found that the unstable *tert*-

butyl hypochlorite can be successfully replaced by stable, commercially available 1,3-dichloro-5,5-dimethylhydantoin or dichloroisocyanuric acid sodium salt [33]. In these reagents, both chlorines are available thereby ensuring maximum efficiency.

2.2
Chiral Ligand and Solvent

The outcome of the AA reaction not only depends on the nitrogen source but also on the solvent and the type of ligand. The sense of enantiofacial selectivity in the AA reaction is determined by the alkaloid moiety of the ligand and is the same as in the AD processes. Thus, the mnemonic device from the AD reaction (see *Comprehensive Asymmetric Catalysis*, Vol. II, p. 746) can be used to predict the enantiofacial selectivity of the AA reaction. The ligand also has an important effect on the regioselectivity in the AA. For example, in the Chloramine-T-based AA of methyl cinnamate the regioselectivity in favor of the benzylic amine was increased from 2:1 to >5:1 under the influence of Cinchona alkaloid ligands [10]. The choice of the ligand spacer also greatly influences the sense of regioselectivity. In the carbamate-AA and amide-AA of cinnamates and styrenes, formation of the benzylic amine was favored under the influence of the PHAL ligand, while the benzylic alcohol was favored when the AQN ligand was employed [20, 22, 34]. In a similar way, the AA of α,β-unsaturated aryl esters with the AQN ligand gave the α-amino isomer, while the PHAL ligand favored the β-amino isomer [35] (Table 2). Interestingly, in the AA of aliphatic α,β-unsaturated alkyl esters the same regioisomer was obtained regardless of the ligand spacer [10, 36].

Additionally, the regioselectivity is strongly solvent-dependent. The most common solvents used in the AA process are alcohols or acetonitrile with high water content (up to 50%). A large amount of water is necessary to ensure high catalytic turnover by enhancing the hydrolysis step in the catalytic cycle. However, in the Boc-AA the opposite was true. Lower water content was favorable by reducing the competitive hydrolysis of the imido osmium species [20], which

Table 2 Reversal of regioselectivity with aryl esters

R	Ligand	Ratio (A:B)	ee (%)
Et	(DHQ)$_2$AQN	1:0	83
4-Cl(C$_6$H$_4$)	(DHQ)$_2$AQN	1:5	89
4-Cl(C$_6$H$_4$)	(DHQ)$_2$PHAL	2.5:1	–

Table 3 The influence of the solvent and the ligand type on the AA reaction

benzylic amide **A**   benzylic alcohol **B**

R	R^1	R^2	R^3	Solvent, ligand	A:B	ee of A, B (%)	Yield (%)
Ac	MeO	H	H	nPrOH/H_2O, (DHQD)$_2$PHAL	2.5:1	96, 62	83
Ac	MeO	H	H	MeCN/H_2O, (DHQD)$_2$PHAL	1:2.4	85, 84	76
Ac	MeO	H	H	MeCN/H_2O, (DHQD)$_2$AQN	1:9	–, 86	58
Cbz	BnO	H	H	nPrOH/H_2O, (DHQ)$_2$PHAL	7.3:1	97, –	76
Cbz	BnO	H	H	MeCN/H_2O, (DHQ)$_2$PHAL	3:1	–	–
Cbz	BnO	H	H	nPrOH/H_2O, (DHQ)$_2$AQN	1:2	–	–
Cbz	BnO	H	H	MeCN/H_2O, (DHQ)$_2$AQN	1:3	–	–
Boc	BnO	BnO	CO$_2$Me	nPrOH/H_2O, (DHQD)$_2$AQN	1:4.5	–, 86	76
Boc	BnO	BnO	CO$_2$Me	MeCN/H_2O, (DHQD)$_2$AQN	1:2.7	–, 84	33
Boc	BnO	BnO	CO$_2$Me	tBuOH/H_2O, (DHQD)$_2$AQN	1:4.8	–, 60	22

led to diol by-products. It seems that alcoholic solvent systems are favorable for the introduction of nitrogen at the benzylic position, while acetonitrile enhances the formation of benzylic alcohols (Table 3) [20, 22, 37]. However, exceptions to this pattern have been observed [38]. Unfortunately in the case of the carbamate-AA of styrene, the enantiomeric purity of the benzylic alcohol (80–0% ee) decreased as the regioselective formation of the benzylic alcohol increased [20]. This problem could be overcome by using the amide AA, which combines good regioselectivity and enantioselectivity in the formation of the benzylic alcohols [22]. Again, a large effect of the solvent on regioselectivity was observed. For some substrates acetonitrile has been reported to have a beneficial effect on reaction rates and helps to suppress diol formation [17, 20, 39], while in other examples the opposite was reported [40].

Lately, Hergenrother and coworkers investigated the influence of the pH on the regioselectivity and chemoselectivity of the carbamate-AA of styrenes [41]. They found that improved regioselectivity (up to 21:1 in favor of the benzylic alcohol), enantioselectivity (up to 74% ee), and chemoselectivity (5% of the diol) were obtained when the reaction was buffered to pH in the range 7.5–8.5.

2.3
Substrate

The magnitude of enantioselectivity and regioselectivity is highly depending on the alkene substitution pattern. The AA works well with three classes of alkenes: monosubstituted alkenes, *cis*-disubstituted alkenes, and *trans*-disubstituted alkenes. Alkenes with aryl and/or carboxylate substituents generally give good to excellent ees (80–99% ee), alkyl substituents on the other side are less favorable (40–80% ee). 1,1-Disubstituted [42] and trisubstituted [43] alkenes give, with some exceptions [44], low enantioselectivities, while tetrasubstituted alkenes does not give sufficient turnover [45].

Janda and coworkers reported the most comprehensive study of the effect of substrate substitution on the regioselectivity of the amide-AA with PHAL ligands [46]. In structurally similar alkenes, the effect of modifying the steric demand and the ligands binding properties on the regioselectivity of the AA was examined. The results were rationalized by a working model for the catalytically active AcN=OsO$_3$–(DHQD)$_2$PHAL complex (Fig. 1) related to Corey's model for the AD reaction. The AA of α,β-unsaturated esters showed an electronic effect on regioselectivity. The more nucleophilic nitrogen ligand of the osmium catalyst prefers to add to the more electrophilic β-carbon of the polarized double bond (Table 4, entries 1–5) [45, 46]. This addition mode may also be explained by hydrophilic interaction of the polar ester group with the solvent instead of the hydrophobic binding pocket. Sterically demanding substituents prefer to point out of the binding pocket of the catalyst. Hence, addition of the oxygen occurs at the carbon near the large group (Table 4, entries 6 and 7; Table 5). However, aromatic substituents are affected by attractive aryl–aryl inter-

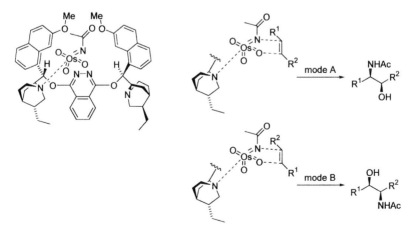

Fig. 1 Model of the AcN=OsO$_3$–(DHQD)$_2$PHAL complex and different alkene binding modes

Table 4 Effect of substrate substitution on regioselectivity in α,β-unsaturated esters[a]

Entry	R[1]	R[2]	Mode[b]	Ratio (A:B)
1	*H*	CO_2Et	A	15.2:1
2	*Me*	CO_2Et	A	1.4:1
3	*4-MeO(C6H4)C(O)OCH2*	CO_2Et	A	>20:1
4	*PhOCH2*	CO_2Et	A	>20:1
5	*2-NaphCH_2OCH_2*	CO_2Et	A	4.3:1
6	TBDPSOCH_2	*CO2(4 McOBn)*	B	1:6.0
7	TBDPSOCH_2	*2-NaphCH2OC(O)*	B	1:17.0

[a] Italics indicate the substituent positioned inside the binding pocket
[b] See Fig. 1

Table 5 Effect of substrate substitution on regioselectivity in homoallylic alcohols[a]

Entry	R[1]	R[2]	Mode[b]	Ratio (A:B)
1	*H*	$TBDPSO(CH_2)_2$	A	>20:1
2	*Et*	$TBDPSO(CH_2)_2$	A	2.0:1
3	*H*	$4-MeO(C_6H_4)O(CH_2)_2$	A	1.2:1
4	Me	*4-MeO(C6H4)O(CH2)2*	B	1:3.2
5	TBDPSOCH_2	*4-MeO(C6H4)C(O)OCH2*	B	11.9:1
6	TBDPSOCH_2	*2-NaphOCH2*	B	>20:1
7	TBDPSOCH_2	*2-NaphCH2OCH2*	B	3.0:1

[a] Italics indicate the substituent positioned inside the binding pocket
[b] See Fig. 1

actions inside the binding pocket leading to addition of the nitrogen at the carbon near to the aromatic group (Table 4, entries 3–7; Table 5, entries 4–7). None of these selection modes were strong enough to completely suppress the others, though careful selection of the alkene substituent led to high regioselectivity by cooperative influence of steric, electronic, and shape complementarity. Certainly, the investigation of steric, electronic, and shape complementarity in the AA with AQN ligands, which are known to provide reversed regioselectivity for certain substrates (see above), would be an interesting extension of this work.

Ojima and coworkers investigated the effect of electronic properties of aryl substituents on the regioselectivity and enantioselectivity in amide-AA reactions of O-substituted 4-hydroxy-2-butenoates with PHAL ligands (Table 6) [47]. For a series of 4-benzoyloxybutenoates, **22**, the regioselectivity of the AA reaction *decreased* with increasing electron-withdrawing ability of the substituent R′. With 4-benzyloxybutenoates, **23**, the dependency of the regioselectivity on the aryl substituent was reversed (i.e., the regioselectivity of the AA reaction *increased* with increasing electron-withdrawing ability of the R′ group). An attractive aromatic–aromatic dipole interaction of the polarized aryl groups with one or other of the two 4-methoxyquinoline moieties of the ligand, which have

Table 6 Effect of aryl substitution on enantioselectivity and regioselectivity in the AA of O-substituted 4-hydroxy-2-butenoates

Entry	R'	Ratio (22A:22B)	%ee of 22A	Ratio (23A:23B)	%ee of 23A
1	MeO	25:1	96	1.2:1	78
2	Me	12:1	96	–	–
3	H	9:1	96	1.3:1	75
4	Br	6.6:1	89	2.9:1	69
5	NO$_2$	4:1	90	11:1	90

opposing dipoles, was proposed to explain the opposite trends. The benzoyloxy groups of **22** were assumed to interact with the 4-methoxyquinoline ring to the left while the benzyloxy groups of **23** were proposed to align with the 4-methoxyquinoline ring to the right (Fig. 1). The enantioselectivities did depend on the electron-withdrawing ability of the aryl substituents in a similar way as the regioselectivities, albeit to a much lesser extent. In a study of the AA of aliphatic α,β-unsaturated aryl esters, Panek and coworkers found a more pronounced dependency of the enantioselectivity on the electron-withdrawing ability of aryl substituents [35].

2.4
Supported Ligands

Solid-supported ligands provide an easy means of recycling the expensive Cinchona alkaloids. Until now, the immobilized Cinchona alkaloid ligands used in the AA process have been attached to different solid supports at the DHQ or DHQD moiety.

The first AA reaction using immobilized chiral ligands was reported in 1998 [48]. A silica-supported ligand **14a** [49] (see Chap. 20.1, Fig. 4) was used in heterogeneous AA reactions of cinnamate derivatives. While heterogeneous amide-AA reactions provided the benzylic amide in enantioselectivities and yield comparable to homogeneous reactions [22], the heterogeneous carbamate-AA reactions were inferior to their homogeneous counterparts [18] with respect to enantioselectivity, yield, and reaction time (Table 7). The polymer-supported

Table 7 Comparison of heterogeneous and homogeneous chiral ligands in the AA reaction

R[1]	R[2]	Ligand		N source (R)	Yield (%), ee (%), t (h)	
		Het	Homo		Het	Homo
H	Me	14a	(DHQ)$_2$PHAL	CO$_2$Et	40, 88, 12	78, 99, 0.8
OMe	Me	14a	(DHQ)$_2$PHAL	CO$_2$Et	43, 92, 12	70, 98, 0.5
NO$_2$	Me	14a	(DHQ)$_2$PHAL	CO$_2$Et	52, 92, 12	75, 97, 1
H	iPr	14a	(DHQ)$_2$PHAL	Ac	71, >99, 7	81, 99, –
OMe	iPr	14a	(DHQ)$_2$PHAL	Ac	76, >99, 9	71, 99, –
H	Me	24	(DHQ)$_2$PHAL	Ts	58, 53, 5	64, 81, 3
H	iPr	25	(DHQ)$_2$PHAL	Ms	92, 87, 24	89, 96, 24

ligands **24** [50] and **25** [51] gave enantioselectivities, yield, and rates inferior to homogeneous analogues [10] in the sulfonamide-AA of cinnamates (Table 7).

3
On the Mechanism of Catalysis in the AA Process

The mechanistic proposals for the AA process are closely related to the AD. In analogy to the AD process, two catalytic cycles have been suggested to be operating (Fig. 2) [17]. The primary cycle incorporates the chiral ligand, which has been found to increase the rate, improve the regioselectivity and chemoselectivity, and induce excellent enantioselectivity. The competing secondary cycle does not involve the chiral ligand and should, therefore, be suppressed. Both cycles involve the trioxoimidoosmium(VIII) complex **26**, which either undergoes hydrolysis (step h) to follow the desired primary cycle or a second addition of an alkene (step a$_2$) to enter into the undesired secondary cycle. In contrast to the AD process, the AA has been carried out under homogeneous conditions

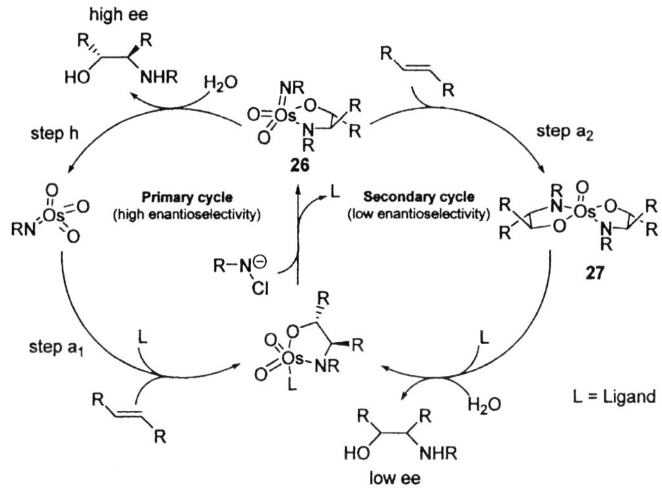

Fig. 2 The two catalytic cycles proposed for the AA reaction

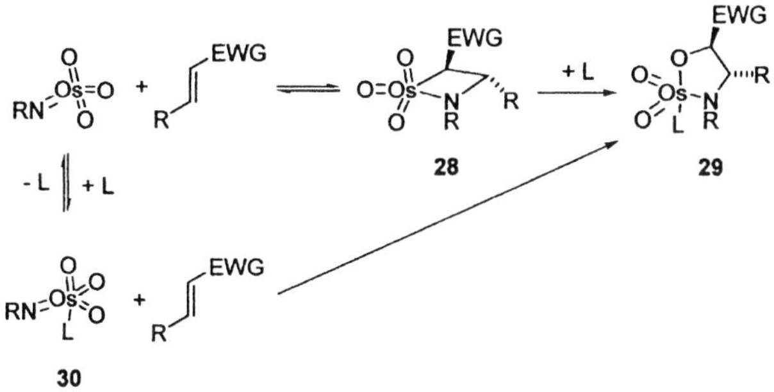

Scheme 2

and, therefore, relies on effective hydrolysis of **26** to avoid the secondary cycle. Sharpless and coworkers discovered that a large water content in the solvent favored the primary over the secondary cycle and increased the overall turnover of the reaction (see above), indicating that hydrolysis of **26** is the rate-limiting step. Additionally, the nature of the nitrogen substituent influences the reaction: large and hydrophobic groups retard hydrolysis, while small substituents lead to improved enantioselectivity, regioselectivity, and chemoselectivity (see above).

As for the mechanistically related AD reaction, two pathways have been suggested for the alkene addition step (step a_1). Sharpless and coworkers proposed a two-step mechanism via an osmaazetidine **28**, followed by ligand-assisted rearrangement to the osmium azaglycolate **29** (Scheme 2) [12, 45] analogous to the proposal for the AD reaction. Janda and coworkers based their studies on an alternative mechanistic proposal, analogous to Criegee and Coreys mechanism for the AD, involving a formal [3+2] cycloaddition of the alkene to the ligand-bound complex **30** (Scheme 2) [46].

Table 8 Asymmetric aminohydroxylation of styrene and methyl cinnamate proceeding in the second catalytic cycle

R	Ligand 4			Ligand 5		
	Ratio (**A:B**)	ee (%), abs. conf.		Ratio (**A:B**)	ee (%), abs. conf.	
		A	B		A	B
H	1:2	48, (*S*)	55, (*R*)	1:2	47, (*S*)	38, (*R*)
CO$_2$Me	1:2	59, (2*S*,3*R*)	25, (2*R*,3*S*)	1:2	–	–

As described for the AD process (see Chap. 20.1), second-cycle AA reactions were induced by the chiral ligands **4** and **5** (Table 8) [52]. These ligands provided good yield (75–93%), but low regioselectivity (1:2) in favor of the benzylic alcohol **B**. Aminohydroxylation of styrene with ligand **4** gave the regioisomers **A** and **B** with 48 and 55% ee, respectively. Methyl cinnamates were converted to the corresponding aminoalcohols with 59 and 25% ee for the regioisomers **A** and **B**, respectively.

Other examples of second-cycle aminohydroxylations involve alkenes containing either anionic groups (e.g., carboxylates, phosphonates, and sulfonates) or cationic groups (e.g., quaternary ammonium) in proximity to the double bond [53]. These substrates undergo rapid and nearly quantitative racemic aminohydroxylations with very low catalyst loadings and only one equivalent of the nitrogen source. The reaction is independent of the chiral ligand indicating that the second cycle is involved. The unusual rate enhancement was explained be the ability of the ionic groups to increase the rate-limiting hydrolysis of the osmium bis(azaglycolate) **27** (Fig. 2).

4
Investigations of Chemoselectivity

A frequently observed side reaction in the AA process is the formation of the diol instead of the aminoalcohol. Generally, the Cinchona alkaloid ligands have a beneficial effect on the chemoselectivity of the AA. However, some examples of AAs producing a considerable amount of diol have been reported in the literature [40, 54, 55].

Lohray and coworkers investigated the formation of diol under AA conditions [56]. IR analyses of a mixture of osmium tetroxide and chloramine-T showed no signs of the formation of the imido osmium compound **29**. Further, no aminoalcohol was observed in the reaction of stoichiometric amounts of osmium tetroxide, Chloramine-T, and chiral ligand with isopropyl cinnamate. Instead the diol was produced with high enantioselectivity (98%). These findings indicated that **29** was not directly formed by a reaction of the osmium tetroxide and the nitrogen source. Lohray and coworkers suggested an alternative mechanism (Scheme 3) involving the addition of the alkene to osmium tetroxide as the initial step in the formation of **29**. Oxidation of the osmium glycolate **1** by the nitrogen source gives an imido glycolate complex **28**, which provides the diol and the imido osmium compound **29** after hydrolysis.

This assumption was supported by the findings that substoichiometric amounts (10 mol%) of osmium tetroxide led to nearly equivalent amounts of the diol (12–14%) (Table 9, entries 1–3). Moreover, the formation of diol was reduced to 7% when the osmate ester **30** (Fig. 3) was used as catalyst. However, the

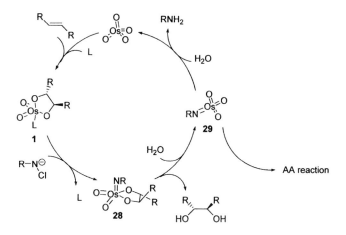

Scheme 3

Table 9 Asymmetric aminohydroxylation of isopropyl cinnamate with 10 mol% osmium tetroxide

Entry	Nitrogen source	Ratio (%), ee (%)		
		Alkene	Diol	Aminoalcohol
1	TsNNaCl	–	14, 82	86, 90
2	MsNNaCl	–	14, 95	86, 94
3	AcNHBr	–	12, –	88, >96
4	CH$_2$ClCOHNBr	13	52, 77	35, 66
5	CHCl$_2$COHNBr	76	12, 71	12, nd
6	CCl$_3$COHNBr	75	25, 26	–
7	tBuCONHBr	86	14, 26	–
8	MeOCH$_2$CONHBr	36	18, 76	46, 83

*nd*not determined

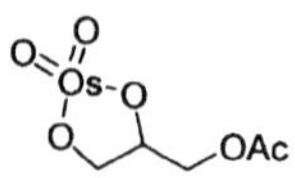

30

Fig. 3

presence of the residual amount of diol (7%) indicates that there must be another route for diol formation. Two alternative routes for diol formation have been suggested by Lohray and coworkers: either the hydrolysis of **29** to osmium tetroxide or the reaction of two oxo groups rather than the imido and one oxo group of **29**. In any case, the formation of diol was found to be strongly dependent on the *N*-substituent of the nitrogen source (Table 9, entries 4–8). Electron-withdrawing and bulky substituents increased the formation of diol and decreased the reaction rate. In an earlier study, Sharpless and coworkers suggested that hydrolysis of **29** accounted for the increased amount of diol in the *tert*-butyl carbamate-AA of styrenes and improved the reaction by reducing the water content of the solvent [20].

Chemoselectivity in the AA process is also dependent on the pH [41] (see above) and the concentration [57] of the reaction mixture. Wuts and coworkers discovered that diol formation increased significantly from 4% to 45% when the concentration was increased from 0.014 to 0.050 g mL^{-1}. Addition of acetamide (1 equivalent) at 0.05 g mL^{-1} improved the diol/aminoalcohol ratio to 5:95. Interestingly, other amides such as trifluoroacetamide and urea inhibited the reaction while the addition of methanesulfonamide led to large quantities of the diol (70%).

5
Synthetic Applications of the Chiral Aminoalcohols

5.1
Cinnamate, Arylacrylates, and Related Compounds

The first application of the AA process was the short and efficient synthesis of the paclitaxel (Taxol) side chain **32** (Scheme 4) [58]. The hydroxysulfonamide **31** crystallized directly from the reaction mixture (69% yield, 82% ee) and was transformed to **32** in two steps. The procedure was further improved when the acetamide-AA was discovered [22]. The hydroxyactamide **33** was easily obtained in good yield (71%) and excellent ee (99% ee) and removal of the *N*-protecting group was nearly quantitative. A direct benzamide-AA gave **34**, the isopropyl ester of **32**, in a single step, though with lower yield (46%) than the acetamide-AA [24].

The AA has also been used in the preparation of analogues of the paclitaxel side chain **32** like protected α,β-diamino acids **35a** and **35b** from **33** [59, 60], β-amino-α-mercapto acids **36a** and **36b** from **33** [59, 61], derivatives bearing a ni-

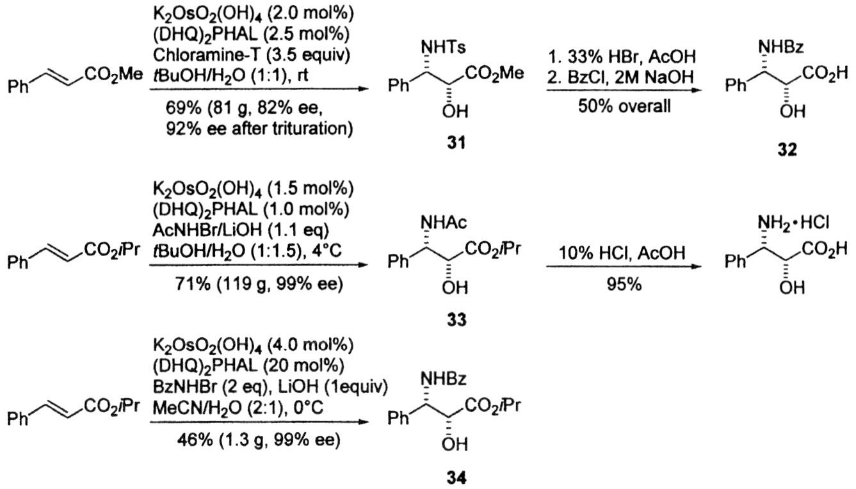

Scheme 4

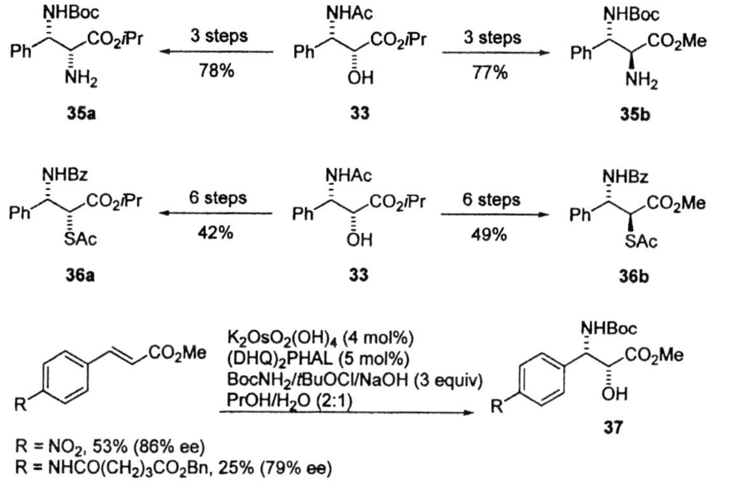

Scheme 5

R = NO$_2$, 53% (86% ee)
R = NHCO(CH$_2$)$_3$CO$_2$Bn, 25% (79% ee)

trogen atom on the phenyl group **37** [62] (Scheme 5), and heteroaromatic derivatives **38a–g** [63–65] (Table 10).

In addition, the AA has been successfully applied to unsaturated phosphonates (Fig. 4) [39, 66] and naphthylacrylate [67].

The AA of cinnamate substrates also provided access to the hydroxyaspargine **40**, a key subunit of ramoplanin A2 [68], and has been used in the synthesis of aziridino alcohol **41** for use as a chiral ligand [69] (Scheme 6).

The AA with reversed regioselectivity (see above) of arylacrylates and heteroaromatic derivatives has been used in a number of applications. The AA of

Table 10 Asymmetric aminohydroxylation of heteroaromatic arylacrylates

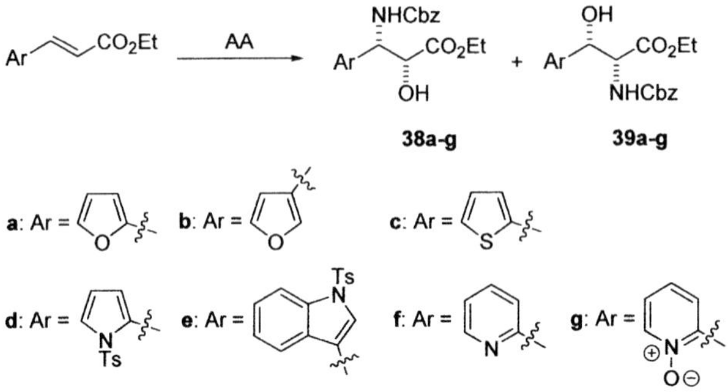

	a: Ar =	b: Ar =	c: Ar =
	d: Ar =	e: Ar =	f: Ar =	g: Ar =

Ar	Ratio (38:39)	ee of 38 (%)	Yield (%)	Reaction conditions
a	>15:1	92	56	$K_2OsO_2(OH)_4$ (4 mol%), $(DHQ)_2PHAL$
b	>15:1	99	45	(5 mol%), nPrOH/H_2O (1:1), CbzNH$_2$/
c	4.8:1	94	67	tBuOCl/NaOH (3.1 equiv)
d	–	–	0	
e	>15:1	93	64	
f	–	–	0	
g	2.3:1	79	80	
a	7:1	87	62	$K_2OsO_2(OH)_4$(4 mol%), $(DHQ)_2PHAL$
c	>20:1	99	71	(5 mol%), tBuOH/H_2O (1:1), CbzNH$_2$/
d	–	–	0	tBuOCl/NaOH (1.2 equiv)
a	7:1	86	41	$K_2OsO_2(OH)_4$(4 mol%), $(DHQ)_2PHAL$ (6 mol%), tBuOH/H_2O (2:1), CbzNH$_2$/ tBuOCl/NaOH (1.1 equiv)

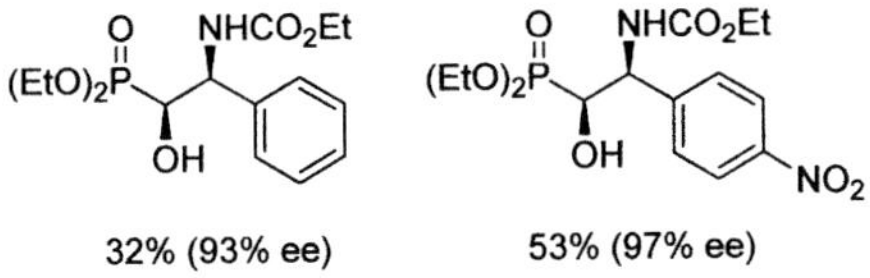

32% (93% ee)   53% (97% ee)

Fig. 4

64% (>99% ee)   **40**

51% (97% ee after recryst.)   **41**

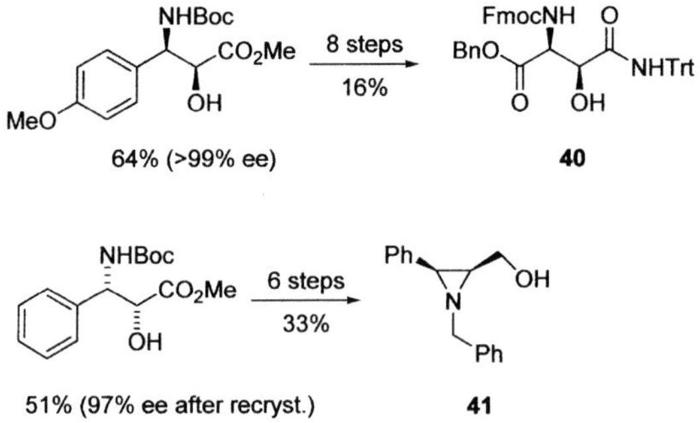

Scheme 6

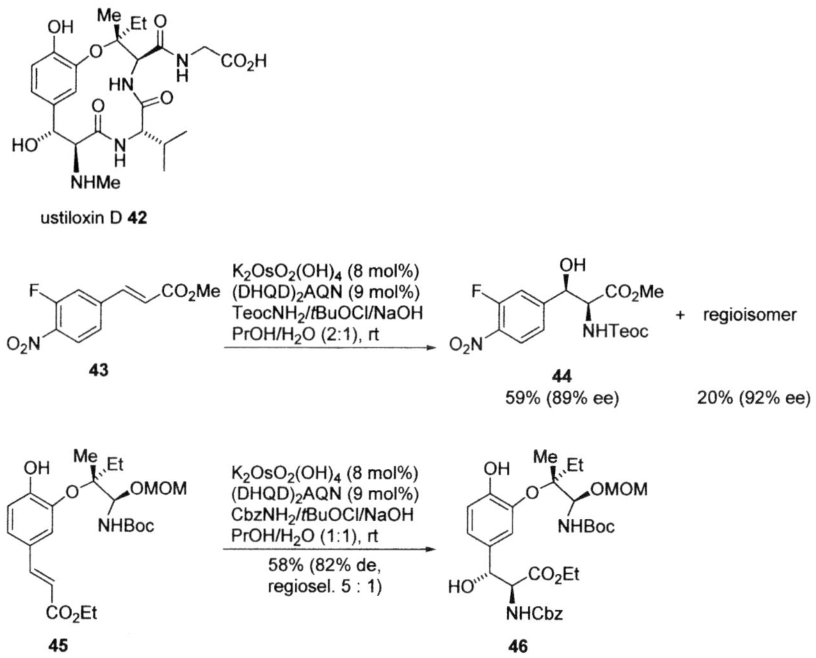

Scheme 7

nitrocinnamate **43** was investigated in the synthesis of **44**, a possible key component in an approach to ustiloxin D (**42**) [38]. Even though the first approach failed, the AA of **45** to **46** was eventually used to prepare the same structural element in the total synthesis of **42** [70] (Scheme 7).

Fluorinated dihydroxyphenyl serine derivatives **48a** and **48b**, potential α- and β-adrenergic receptor antagonists, have been prepared from the corresponding fluorinated cinnamates **47a** and **47b** by the (DHQD)$_2$AQN mediated AA [37] (Scheme 8). Reactions in propanol/water (1:1) gave the best results with respect to yield (76% and 38%) and enantioselectivity (86% and 82% ee).

In an attempt to synthesize a component **50** of exochelin MN (**51**), the regioinversed AA of imidazol acrylates **49** under the influence of the (DHQD)$_2$AQN ligand was investigated (Scheme 9). While hydroxyphenyl-substituted substrates such as **48a** and **48b** gave satisfactory yields (45–55%), regioselectivities (3:1–1:0), and enantioselectivities (84–86% ee), no reaction was observed with imidazole-substituted substrates like **48c** [71]. However, the synthesis of the indole-substituted amino acid **52**, a component of cyclomarin A (**53**), was achieved in 36% yield (Scheme 9) [72]. The diastereoselectivity of the AA reaction was determined to be 95:5.

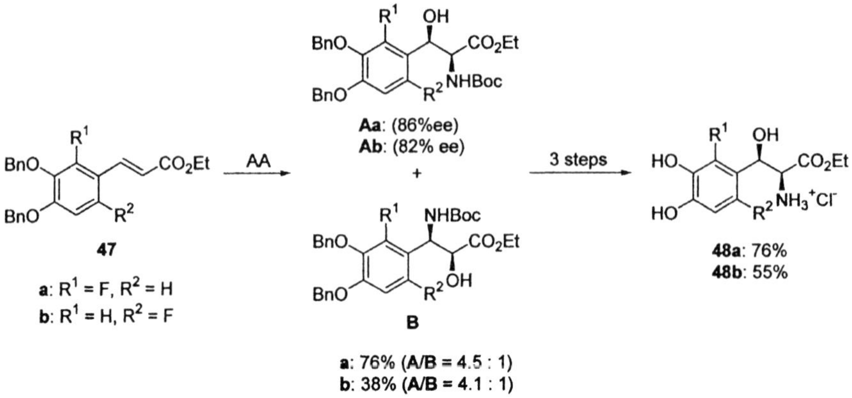

Scheme 8

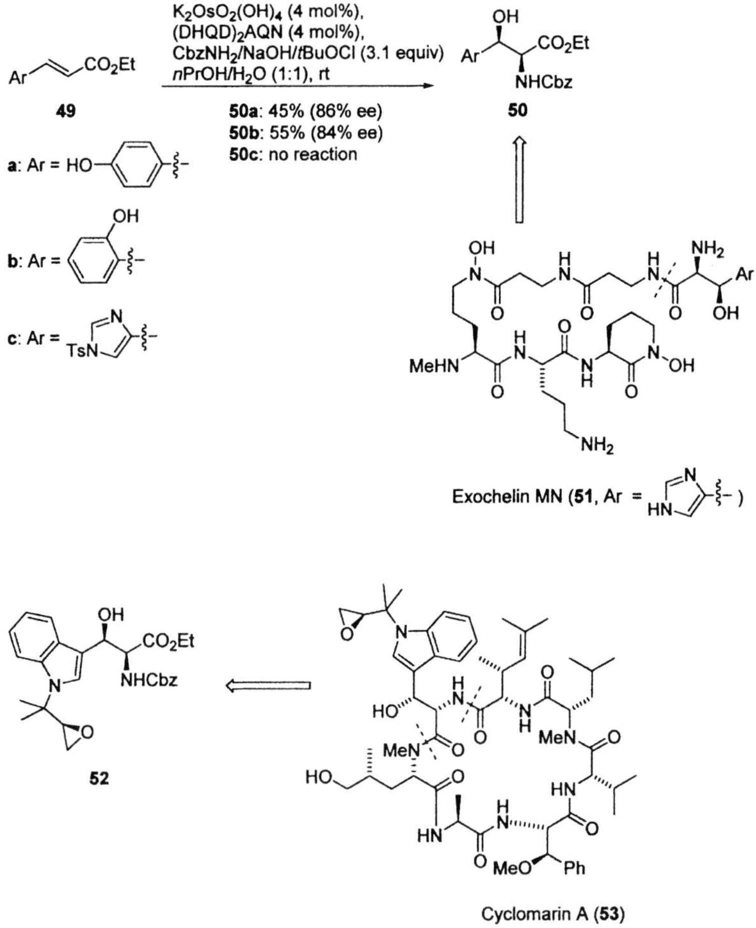

Exochelin MN (**51**, Ar = imidazole)

Cyclomarin A (**53**)

Scheme 9

5.2
Styrenes and Related Compounds

The regioselective and enantioselective AA of styrenes [20, 21, 26, 73] provides an efficient and selective route to arylglycines, which are found in a wide range of bioactive compound such as amoxicillins, nocardicins, cephalecins, and glycopeptide antibiotics (e.g., vancomycin). The AA of a number of substituted styrenes was investigated to examine the effect of the substituent on the regioselectivity [20]. Subsequent oxidation of the crude mixture of regioisomers afforded the arylglycines **54**, which were easily separated from the amino ketones **55** (Scheme 10). Arylglycinols **56–61** (Fig. 5), prepared by the AA reaction, were transformed to the corresponding arylglycines, components in the syntheses of the vancomycin CD and DE ring systems (from **56**) [74], conformationally restricted L-arginine derivatives as probes for the binding with nitric oxide synthase (from **57** and **58**) [75, 76], and the DEF ring system of ristocetin A (from **59**) [77] and teicoplanin (from **60** and **61**) [78].

Additionally, arylglycinol **62** was transformed to the corresponding cyclohexylglycine **63** (Scheme 11) [79]. Furthermore, arylglycinols obtained by the AA of styrene derivatives were applied in the synthesis of chiral bis(oxazoline) **64** [80] and oxazolidin-2-ones such as **65** (Scheme 11) [81]. Oxazolidin-2-ones were also provided by a practical one-pot synthesis utilizing a modified AA procedure with urethane as nitrogen source [33].

Heteroaromatic-substituted vinyl compounds were successfully subjected to the AA reaction, too. In contrast to styrene substrates, 2-vinylfuran was convert-

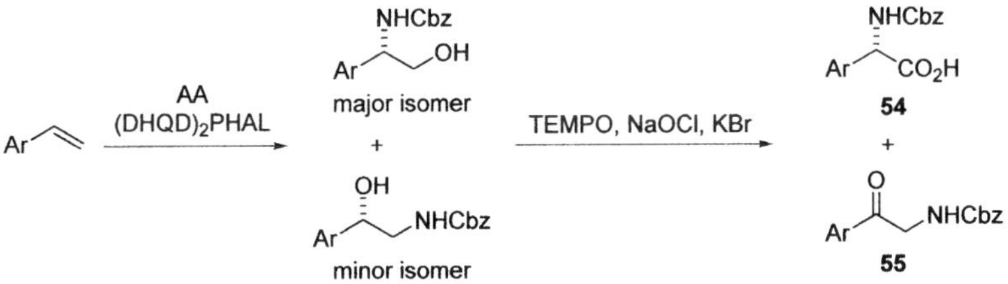

Scheme 10

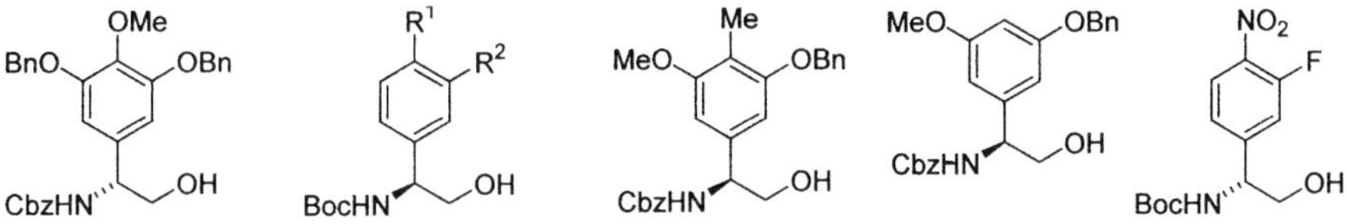

56
68% (96% ee)

57: R^1 = NHCbz, R^2 = H
59% (75% ee)
58: R^1 = H, R^2 = NHCbz
66% (96% ee)

59
66% (>90% ee)

60
78% (>99% ee)

61
75% (97% ee)

Fig. 5

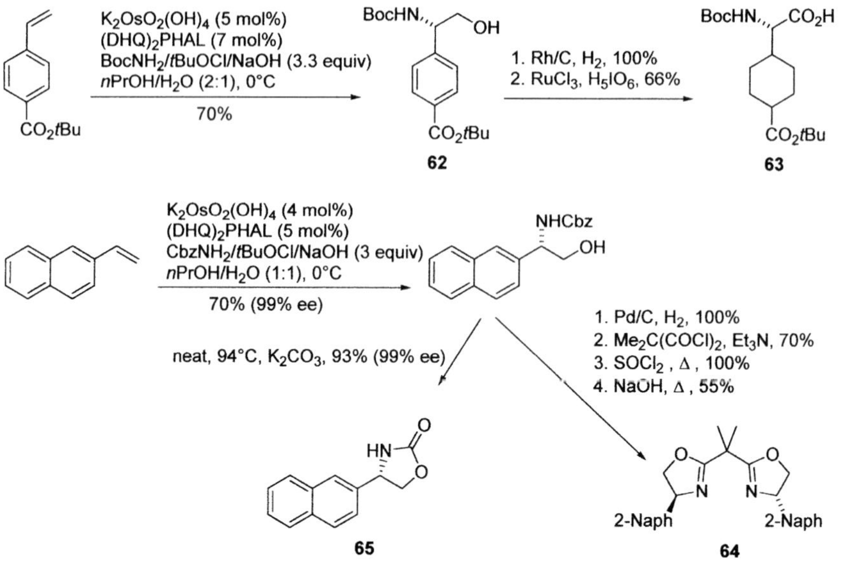

Scheme 11

ed to the aminoalcohols **66** with low regioselectivity (1:2) in favor of the secondary alcohol **66B**. The minor regioisomer **66A** was obtained in 86% ee, while the major isomer **66B** was formed in much lower enantiomeric excess (14% ee) [63]. The aminoalcohols were separated as their corresponding trimethylsilyl ethers and the minor isomer was transformed into deoxymannojirimycin and deoxygulonojirimycin (Scheme 12) [82]. The AA of vinylindoles was investigated in an approach towards bisindole alkaloids. Vinylindoles were converted to the aminoalcohols with the same sense of regioselectivity as observed in the AA of styrenes. Boc carbamate as nitrogen source was superior to Cbz carbamate with respect to regioselectivity and enantioselectivity. Aminoalcohol **67A**, obtained from 6-bromovinylindole in 65% yield and 94% ee, was a key intermediate in the syntheses of hamacanthin A and dragmacidin A (Scheme 12) [83].

5.3
Conjugated Esters

Conjugated esters are a third class of substrates well suited for the AA. The β-amino-α-hydroxy esters **68–71**, resulting from the AA with PHAL ligands, are components of several small bioactive peptides, like the renin inhibitor KRI 1314 [84], the aminopeptidase inhibitors bestatin, amastatin, and microginin [85], and related compounds [86] (Fig. 6). The ethyl ester analogue of cyclohexylnorstatine **68** was further converted to the Abbott amino diol **72**, a core unit of the renin inhibitor zankiren (Fig. 6) [87].

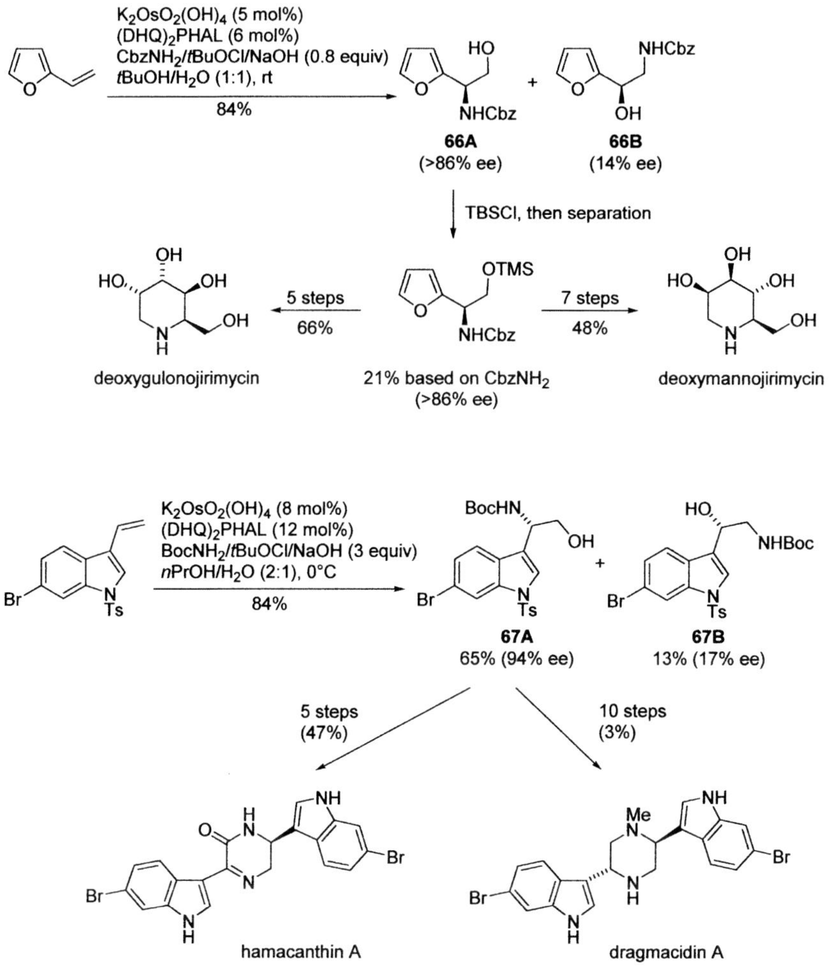

Scheme 12

The β-amino-α-hydroxy esters have been converted to diamino acids [88, 89] and related compounds like the nitrogen-substituted azetidinone shown in Scheme 13, a key structure in the synthesis of the commercially available antibiotic loracarbef [90] by substitution of the alcohol moiety with an azide. Several approaches have been used to achieve this transformation. Mesylation of the alcohol followed by substitution with sodium or trimethylsilyl azide provided *cis*-diamino acids [88, 89]. *trans*-Diamino acids were obtained by ring opening of the aziridine [88] or by inversion of the alcohol bearing carbon followed by substitution under Mitsunobo conditions [90].

The possibility to obtain amino diols from β-amino-α-hydroxy esters (see above) was crucial in the exploration of an AA approach towards D-ango-

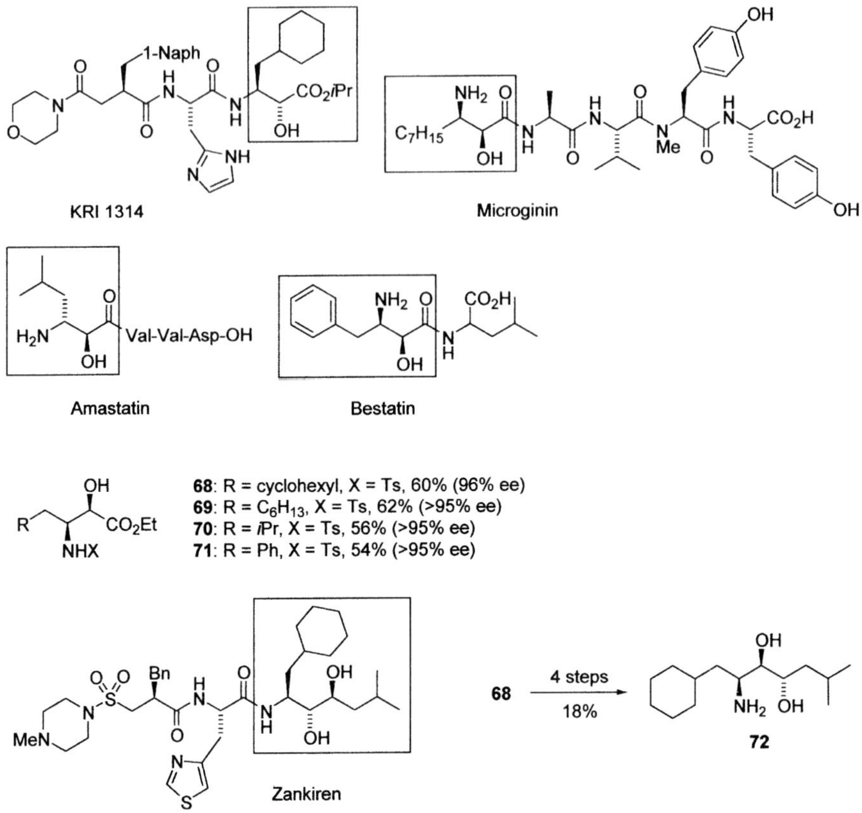

Fig. 6

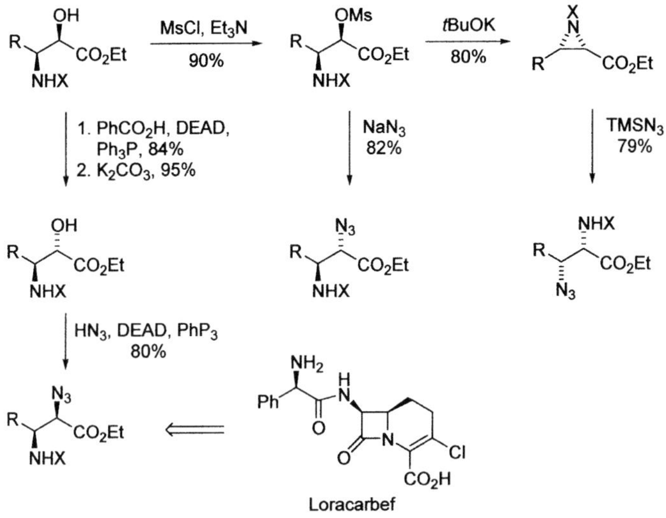

Scheme 13

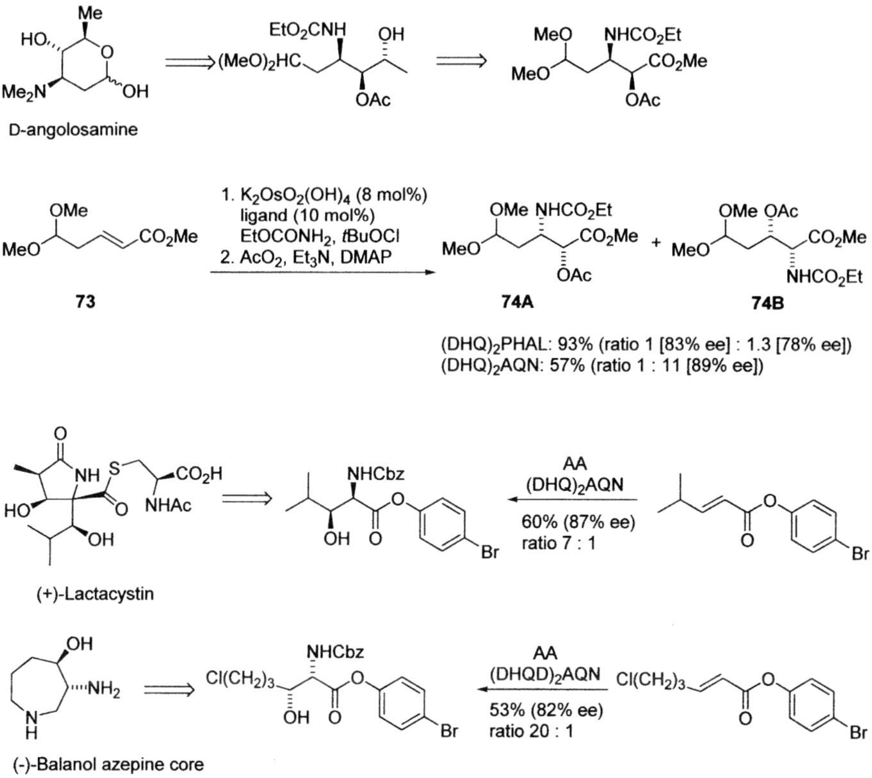

Scheme 14

losamine, the aminosugar component in medermycin [36]. However, the AA of the 5,5-dimethoxy-substituted ester **73** provided the unwanted α-amino-β-hydroxy ester **74B** as the major regioisomer regardless of the chiral ligand (Scheme 14). Panek and coworkers found that good regioselectivity towards the α-amino-β-hydroxy esters was obtained in the AA of conjugated aryl esters (see above and Table 2) [35]. This reverse regioselectivity has been exploited in the synthesis of lactacystin [91, 92], a selective and potent inhibitor of the 20S proteasome, and the azepine core of balanol [93], a potent protein kinase C inhibitor.

5.4
Miscellaneous

The AA of a more unusual AA substrate, the cyclic *cis*-alkene **75**, has been exploited in the short synthesis of (+)-ioline, a pyrrolizidine alkaloid from rye grass and tall fescue (Scheme 15) [40, 94]. Slow addition of the potassium osmate gave the regioisomers **76A** and **76B** in 52% yield. A high ligand-to-osmium

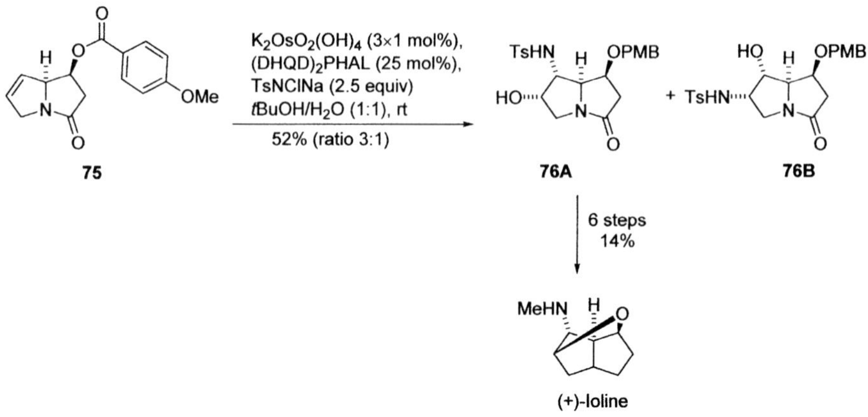

Scheme 15

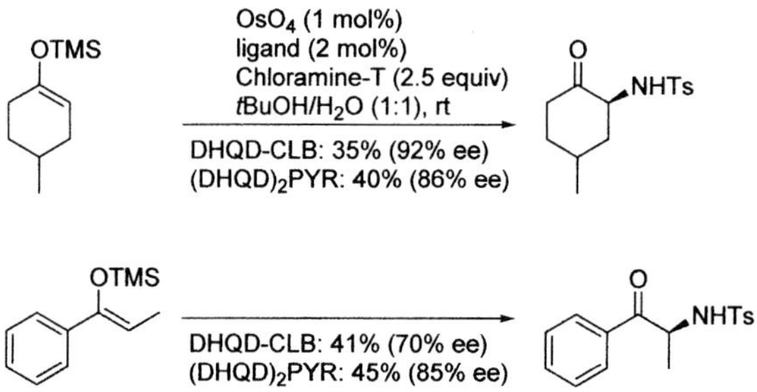

Scheme 16

ratio was necessary to suppress diol formation (21% yield). The diastereomers from reaction at the unfavorable face of the double bond (corresponding to the unfavorable enantiomer in AAs of achiral alkenes) were not reported. Trisubstituted silyl enol ethers are another unusual class of substrates for AA reactions; these were easily converted to α-amino ketones by the AA reaction in moderate yield (28–45%) and good enantioselectivities (70–92% ee) (Scheme 16) [44].

2-Amino-1,2-diphenylethanol, obtained by the AA of stilbene, has been used to introduce chirality in the synthesis of aminocyclohexitols [95] and the 3-amino-β-lactam moiety of loracarbef [96].

Donohoe and coworkers developed a racemic aminohydroxylation using an intramolecular nitrogen source [30, 31]. Treatment of allylic carbamates 77 with *tert*-butyl hypochlorite and sodium hydroxide in the presence of $K_2OsO_2(OH)_4$ and an amine ligand led to the regioselective formation of hydroxyoxazolidinones 78 (Scheme 17). Both acyclic and cyclic allylic carbamates have been employed in this transformation.

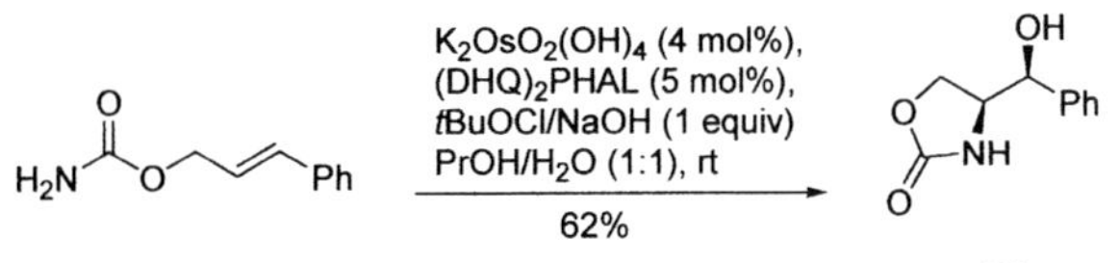

Scheme 17

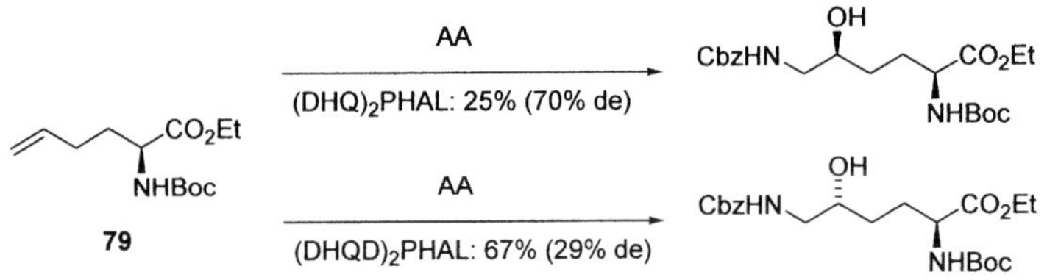

Scheme 18

5.5
Diastereoselectivity in the AA Reaction

A few recent studies have described the AA of substrates containing chiral [40, 70, 72, 94, 97, 98] or prochiral [99–102] centers. In the AA of chiral substrates, double diastereoselectivity arose from the interaction of the substrate with the chiral ligand. The AA proceeded with the sense of facial selectivity expected for the DHQD or DHQ ligands. The effect of the chiral center on the facial selectivity has not been investigated. A study of the AA of the chiral alkene **79** with the pseudoenantiomeric forms of the PHAL ligands revealed that the DHQ and the DHQD ligands led to matched (70% de) and mismatched (29% de) reactions, respectively (Scheme 18) [97].

The aminohydroxylation of racemic Baylis–Hillman alkenes provided regio-isomerically pure, racemic mixtures of the α-hydroxy-β-amino esters, with the *syn* (diol) product as the major diastereomer (Table 11) [97]. The diastereoselectivity increased with increasing size of either the allylic substituent or the ester group. As observed in other studies of related substrates [53, 103–105], neither the rate, the selectivity, nor the yield were noticeably affected when a chiral ligand (e.g., (DHQ)₂PHAL) was added and enantioselective aminohydroxylation could not be obtained.

The most striking example of diastereoselective AA reactions is the desymmetrization of dienylsilanes developed by Landais and coworkers (Table 12) [99–102]. Good regioselectivity and diastereoselectivity but moderate enantioselectivity were achieved in the AA reaction of 2,5-cyclohexadienylsilanes with (DHQ)₂PYR as chiral ligand. The addition proceeded *anti* to the silyl group with the nitrogen α to the silyl-substituted carbon. The enantiomeric purity of **80** (R=OH, 68% ee) was easily improved to >99% ee by a single recrystallization. The allylic alcohol **80** offered an efficient entry to amino carbacycle **81** and ami-

Table 11 Diastereoselective aminohydroxylation of Baylis–Hillman alkenes

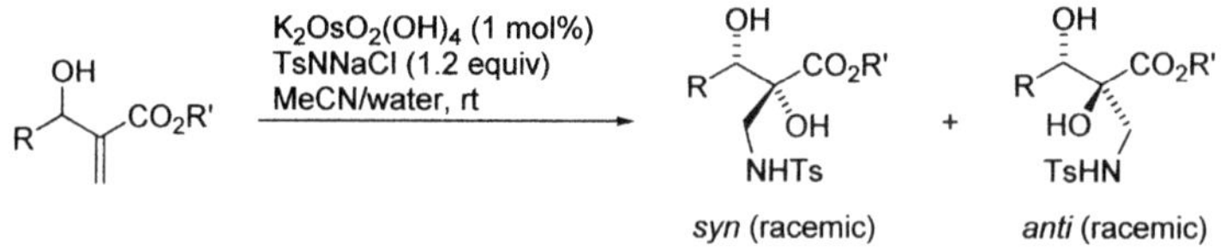

R	R'	Syn:Anti	Yield (%)
Me	Me	88:12	88
Et	Me	90:10	78
iPr	Me	98:2	71
Ph	Me	>99:1	23
Me	Et	86:14	70
Me	iPr	89:11	65

Table 12 AA of 2,5-cyclohexadienylsilanes

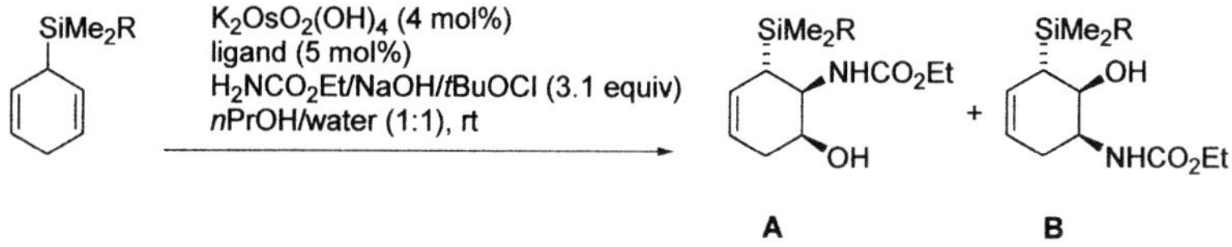

R	Ligand	Ratio (A:B)	ee of A (%)	de (%)	Yield (%)
OH	(DHQ)$_2$PYR	>98:2	68	>98%	75
tBu	(DHQ)$_2$PYR	92:8	81	>98%	65
tBu	Quinuclidine	7:3	–	–	
tBu	iPr$_2$NEt	1:1	–	–	

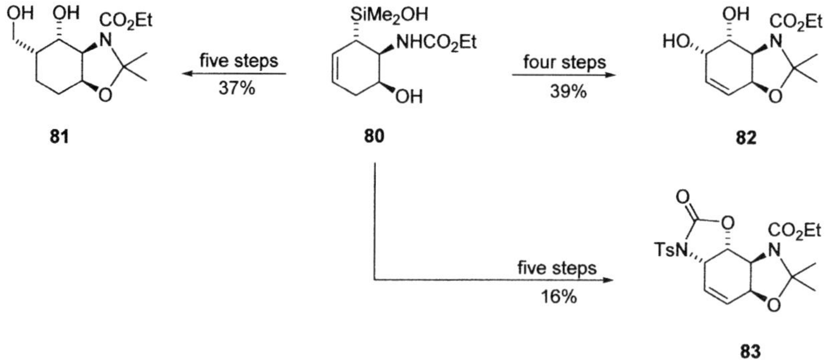

Scheme 19

nocylitols, such as the fortamine precursor **82** and the aglycon moiety of strep-
tomycin antibiotics **83** (Scheme 19).

6
Conclusion

Compared to the closely related AD reaction, the AA reaction offers several new
challenges, for example, control of regioselectivity and chemoselectivity. Highly significant improvements have been made in these areas and AA has developed into a very powerful tool for the asymmetric introduction of the vicinal
aminoalcohol moiety. Today, a wide range of alkenes can be converted into the
corresponding aminoalcohols with excellent enantioselectivity and regioselectivity. In many cases, information about the optimum reaction conditions with
respect to factors like substrate structure can be found in the literature. However, a better understanding and control of factors influencing the regioselectivity
and a general improvement of the AA of difficult substrates are important tasks
for the further development of the AA reaction. The development of new ligands
for an enantioselective second cycle may have a great impact on the further development of the AA reaction.

References

1. Sharpless KB, Patrick DW, Truesdale LK, Biller SA (1975) J Am Chem Soc 97:2305
2. Patrick DW, Truesdale LK, Biller SA, Sharpless KB (1978) J Org Chem 43:2628
3. Sharpless KB, Chong AO, Oshima K (1976) J Org Chem 41:177
4. Herranz E, Biller SA, Sharpless KB (1978) J Am Chem Soc 100:3596
5. Herranz E, Sharpless KB (1978) J Org Chem 43:2544
6. Herranz E, Sharpless KB (1980) J Org Chem 45:2710
7. Herranz E, Sharpless KB (1982) Org Synth 61:85
8. Herranz E, Sharpless KB (1982) Org Synth 61:93
9. Rubinstein H, Svendsen JS (1994) Acta Chem Scand 48:439
10. Li G, Chang H-T, Sharpless KB (1996) Angew Chem Int Ed 35:451
11. Reiser O (1996) Angew Chem Int Ed 35:1308
12. Kolb HC, Sharpless KB (1998) Asymmetric aminohydroxylation. In: Beller M, Bolm C
    (eds) Transition metals for organic synthesis. Wiley-VCH, Weinheim, vol 2, p 243
13. O'Brien P (1999) Angew Chem Int Ed 38:326
14. Bolm C, Hildebrand JP, Muniz K (2000) In: Ojima I (ed) Catalytic asymmetric synthesis.
    Wiley-VCH, New York, p 399
15. Schlingloff G, Sharpless KB (2001) Asymmetric aminohydroxylation. In: Katsuki T (ed)
    Asymmetric oxidation reactions. Oxford UP, Oxford, p 104
16. Bodkin JA, McLeod MD (2002) J Chem Soc Perkin Trans 1 2733
17. Rudolph J, Sennhenn PC, Vlaar CP, Sharpless KB (1996) Angew Chem Int Ed 35:2810
18. Li G, Angert HH, Sharpless KB (1996) Angew Chem Int Ed 35:2813
19. Reddy KL, Dress KR, Sharpless KB (1998) Tetrahedron Lett 39:3667
20. Reddy KL, Sharpless KB (1998) J Am Chem Soc 120:1207
21. O'Brien P, Osborne SA, Parker DD (1998) Tetrahedron Lett 39:4099
22. Bruncko M, Schlinloff G, Sharpless KB (1997) Angew Chem Int Ed 36:1483
23. Demko ZP, Bartsch M, Sharpless KB (2000) Org Lett 2:2221
24. Song CE, Oh CR, Roh EJ, Lee S, Choi JH (1999) Tetrahedron Asymmetry 10:671
25. Goossen LJ, Liu H, Dress KR, Sharpless KB (1999) Angew Chem Int Ed 38:1080

26. O'Brien P, Osborne SA, Parker DD (1998) J Chem Soc Perkin Trans 1 2519
27. Pilcher AS, Yagi H, Jerina DJ (1998) J Am Chem Soc 120:3520
28. Dress KR, Gooßen LJ, Liu H, Jerina DM, Sharpless KB (1998) Tetrahedron Lett 39:7669
29. Gooßen LJ, Liu H, Dress KR, Sharpless KB (1999) Angew Chem In Ed 38:1080
30. Donohoe TJ, Johnson PD, Helliwell M, Keenan M (2001) Chem Commun 2078
31. Donohoe TJ, Johnson PD, Cowley A, Keenan M (2002) J Am Chem Soc 124:12934
32. Teeter HM, Bell EW (1952) Org Synth 32:20; (1963) Coll Vol 4:125
33. Barta NS, Sidler DR, Somerville KB, Weissman SA, Larsen RD, Reider PJ (2000) Org Lett 2:2821
34. Tao B, Schlingloff G, Sharpless KB (1998) Tetrahedron Lett 39:2507
35. Morgan AJ, Masse CE, Panek JS (1999) Org Lett 1:1949
36. Davey RM, Brimble MA, McLeod MD (2000) Tetrahedron Lett 41:5141
37. Kim IH, Kirk KL (2001) Tetrahedron Lett 42:8401
38. Park H, Cao B, Joullié MM (2001) J Org Chem 66:7223
39. Thomas AA, Sharpless KB (1999) J Org Chem 64:8379
40. Blakemore PR, Kim S-K, Schultz VK, White JD, Yokochi AFT (2001) J Chem Soc Perkin Trans 1 1831
41. Nesterenko V, Byers JT, Hergenrother PJ (2003) Org Lett 5:281
42. Avenoza A, Cativiela C, Corzana F, Peregrina JM, Sucunza D, Zurbano MM (2001) Tetrahedron Asymmetry 12:949
43. Clark JS, Townsend RJ, Blake AJ, Teat SJ, Johns A (2001) Tetrahedron Lett 42:3235
44. Phukan P, Sudalai A (1998) Tetrahedron Asymmetry 9:1001
45. Sharpless KB, Guigen L (Scripps Research Inst) (1998) US5767304
46. Han H, Cho C-W, Janda KD (1999) Chem Eur J 5:1565
47. Chuang C-Y, Vassar VC, Ma Z, Geney R, Ojima I (2002) Chirality 14:151
48. Song CE, Oh CR, Lee SW, Lee S, Canali L, Sherrington DC (1998) Chem Commun 2435
49. Song CE, Yang JW, Ha HJ (1997) Tetrahedron Asymmetry 8:841
50. Nandanan E, Phukan P, Pais GCG, Sudalai A (1999) Indian J Chem 38B: 287
51. Mandoli A, Pini D, Agostini A, Salvadori P (2000) Tetrahedron Asymmetry 11:4039
52. Andersson MA, Epple R, Fokin VV, Sharpless KB (2002) Angew Chem Int Ed 41:472
53. Sharpless KB, Fokin V (Scripps Research Inst) (2002) US6350905
54. Sunose M, Anderson KM, Orpen AG, Gallagher T, Macdonald SJF (1998) Tetrahedron Lett 39:8885
55. Baumgartner J, Weber H (2001) Chirality 13:159
56. Lohray BB, Bhushan V, Reddy GJ, Reddy AS (2002) Indian J Chem 41B: 161
57. Wuts PEM, Anderson AM, Goble MP, Mancini SE, VanderRoest RJ (2000) Org Lett 2:2667
58. Li G, Sharpless KB (1996) Acta Chem Scand 50:649
59. Lee S-H, Yoon J, Nakamura K, Lee Y-S (2000) Org Lett 2:1243
60. Lee S-H, Yoon J, Chung S-H, Lee Y-S (2001) Tetrahedron 57:2139
61. Lee S-H, Qi X, Yoon J, Nakamura K, Lee Y-S (2002) Tetrahedron 58:2777
62. Montiel-Smith S, Cervantes-Mejía V, Dubois J, Guénard D, Guéritte F, Sandoval-Ramírez J (2002) Eur J Org Chem 2260
63. Bushey ML, Haukaas MH, O'Doherty GA (1999) J Org Chem 64:2984
64. Raatz D, Innertsberger C, Reiser O (1999) Synlett 1907
65. Zhang H, Xia P, Zhou W (2000) Tetrahedron Asymmetry 11:3439
66. Cravotto G, Giovenzana GB, Pagliarin R, Palmisano G, Sisti M (1998) Tetrahedron Asymmetry 9:745
67. Phukan P, Sudalai A (2000) Indian J Chem 39B: 291
68. Boger DL, Lee RJ, Bounaud P-Y, Meier P (2000) J Org Chem 65:6770
69. Andersson PG, Guijarro D, Tanner D (1997) J Org Chem 62:7364
70. Cao B, Park H, Joullie MM (2002) J Am Chem Soc 124:520
71. Dong L, Miller MJ (2002) J Org Chem 67:4759
72. Sugiyama H, Shioiri T, Yokokawa F (2002) Tetrahedron Lett 43:3489
73. Medina E, Moyano A, Pericàs MA, Riera A (2000) Helv Chim Acta 83:972

74. Boger DL, Borzilleri RM, Nukui S, Beresis RT (1997) J Org Chem 62:4721
75. Atkinson RN, Moore L, Tobin J, King SB (1999) J Org Chem 64:3467
76. Li X, Atkinson RN, King SB (2001) Tetrahedron 57:6557
77. Pearson AJ, Heo J-N (2000) Org Lett 2:2987
78. Boger DL, Kim SH, Mori Y, Weng J-H, Rogel O, Castle SL, McAtee JJ (2001) J Am Chem Soc 123:1862
79. Venkatraman S, Njoroge FG, Girijavallabhan V, McPhail AT (2002) J Org Chem 67:2686
80. Crosignani S, Desimoni G, Faita G, Righetti PP (1998) Tetrahedron 54:15721
81. Li G, Lenington R, Willis S, Kim SH (1998) J Chem Soc Perkin Trans 1 1753
82. Haukaas MH, O'Doherty GA (2001) Org Lett 3:401
83. Yang C-G, Wang J, Tang X-X, Jiang B (2002) Tetrahedron Asymmetry 13:383
84. Upadhya TT, Sudalai A (1997) Tetrahedron Asymmetry 8:3685
85. Phukan P (2002) Indian J Chem 41B: 1015
86. Keding SJ, Dales NA, Lim S, Beaulieu D, Rich DH (1998) Synth Commun 28:4463
87. Chandrasekhar S, Mohapatra S, Yadav JS (1999) Tetrahedron 55:4763
88. Han H, Yoon J, Janda KD (1998) J Org Chem 63:2045
89. Hennings DD, Williams RM (2000) Synthesis 1310
90. Lee J-C, Kim GT, Shim YK, Kang SH (2001) Tetrahedron Lett 42:4519
91. Panek JS, Masse CE (1999) Angew Chem Int Ed 38:1093
92. Masse CE, Morgan AJ, Adams J, Panek JS (2000) Eur J Org Chem 2513
93. Masse CE, Morgan AJ, Panek JS (2000) Org Lett 2:2571
94. Blakemore PR, Schultz VK, White JD (2000) Chem Commun 1263
95. Kim KS, Choi SO, Park JM, Lee YJ, Kim JH (2001) Tetrahedron: Asymmetry 12:2649
96. Palomo C, Ganboa I, Kot A, Dembkowski L (1998) J Org Chem 63:6398
97. Pringle W, Sharpless KB (1999) Tetrahedron Lett 40:5151
98. Löhr B, Orlich S, Kunz H (1999) Synlett 1139
99. Angelaud R, Landais Y (1997) Tetrahedron Lett 38:1407
100. Landais Y (1998) Chemia 52:104
101. Angelaud R, Babot O, Charvat T, Landais Y (1999) J Org Chem 64:9613
102. Rahman NA, Landais Y (2002) Curr Org Chem 6:1369
103. Rubin AE, Sharpless KB (1997) Angew Chem Int Ed 36:2637
104. Gontcharov AV, Liu H, Sharpless KB (1999) Org Lett 1:783
105. Fokin VV, Sharpless KB (2001) Angew Chem Int Ed 40:3455

Supplement to Chapter 24
Allylic Substitution Reactions

Jean-François Paquin, Mark Lautens

Department of Chemistry, University of Toronto,
Toronto, Ontario, M5S 3H6, Canada
e-mail: jfpaquin@chem.utoronto.ca, mlautens@chem.utoronto.ca

Keywords: Allylic substitution, Allylation, Allylic alkylation, π-Allyl complexes, Palladium, Molybdenum, Ruthenium, Iridium

1	**Introduction**	73
2	**Mechanism**	74
3	**Survey of Reactions**	74
3.1	Substrates with Identical Substituents at C1 and C3 (RCH=CH–CHXR)	74
3.1.1	Acyclic Substrates	75
3.1.2	Cyclic Substrates	80
3.2	Substrates with Different Substituents at C1 and C3 (R^1CH=CH–CHR^{2}X)	83
3.3	Substrates with Identical Geminal Substituents at C1 or C3 (R^1CH=CH–C(R^2)$_2$X) or (R^1)$_2$C=CH–CHR^{2}X)	85
3.4	*meso*-Substrates with Two Enantiotopic Leaving Groups	88
3.5	Substrates with Two Geminal Enantiotopic Leaving Groups	89
3.6	Reactions of Prochiral Nucleophiles	90
4	**Conclusion**	91
	References	91

1
Introduction

An enormous amount of work in the area of transition-metal-catalyzed asymmetric allylic substitution has been done over the last few years. Since the publication of the original chapter [1], a number of books [2–5], reviews [6–12], and accounts [13, 14] covering this field have been published. This update will cover the recent developments from 1999 to early 2003.

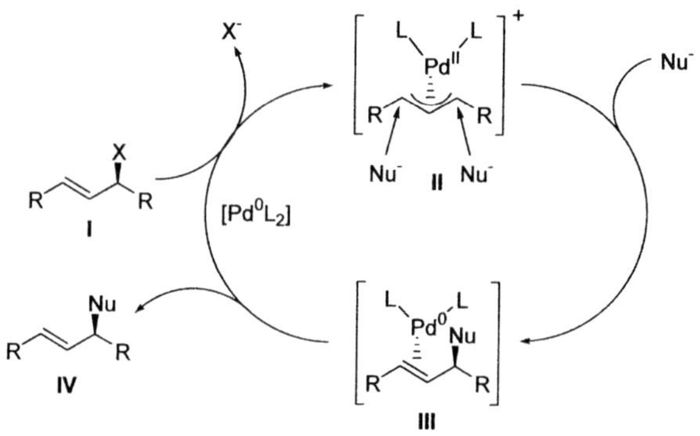

Scheme 1

2
Mechanism

The generally accepted mechanism of the palladium-catalyzed substitution reaction is shown in Scheme 1. For a more detailed discussion, the reader is referred to the original chapter [1]. A number of papers which probe the enantiodetermining step for specific ligands have been published over the past four years but will not be discussed further [15–28]. The effect of catalyst loading [29], the ionization of I with different Pd^0 complexes [30, 31], the mechanism of the η^3–η^1–η^3 isomerization [32] of intermediate II, and the behavior of the olefin–Pd(0) complex III have been studied [33].

3
Survey of Reactions

Each class of substrate and nucleophile that have been reported to successfully react is discussed in this section. For a detailed discussion of rationale for the organization and classification, see the original chapter [1]. The best ligand(s) for each substrate class was(were) selected rather than providing a comprehensive listing of all ligands reported. When possible, ligands which give >90% ee are listed but in the absence of this level of selectivity the best results obtained to date with a specific ligand are presented.

3.1
Substrates with Identical Substituents at C1 and C3 (RCH=CH–CHXR)

This class of substrates has been intensively studied over the past four years and numerous ligands that afford the alkylation product with >90% ee have been reported. The substrate can be either acyclic or cyclic and the difficulty to extrapolate results between these classes is well established. In general, a racemic mixture can be converted to a single product, since the stereochemical information

contained in the substrate is lost in the symmetrical achiral π-allyl intermediate. The regioselectivity of the nucleophilic attack, and hence the enantioselectivity of the process, can be dictated by the chiral ligand.

3.1.1
Acyclic Substrates

As mentioned before, acyclic substrates have been studied in detail, in particu-
lar, the reactions of 1,3-diphenylpropenyl ester (**19**, X=COR) or carbonate (**19**,
X=OCO$_2$R), the "gold standard" for this class. Selected ligands for the allylic
alkylation, amination, and etherification are shown in Fig. 1 and representative
data are presented in Table 1.

Among the more surprising observations, ligand **1** bearing a free alcohol pro-
duces the enantiomeric product to that arising using the benzoate **2** [34]. A similar outcome in which changing a substituent reversed the stereoselectivity was
also reported for a P,N-ligand derived from (S)-valine [52, 53]. Use of trimeth-
ylborate as a nucleophilic source of methanol in conjunction with ligand **17** al-

Table 1

L*	X	Nu	% Yield	% ee (abs config)	Ref.
1	OAc	NaCH(CO$_2$Me)$_2$	98	92 (S)	[34]
2	OAc	NaCH(CO$_2$Me)$_2$	98	90 (R)	[34]
3	OAc	CH$_2$(CO$_2$Me)$_2$/BSA	>99	94–97 (S)	[35, 36]
4	OAc	CH$_2$(CO$_2$Me)$_2$/BSA	>99	97–98 (S)	[37]
5	OAc	CH$_2$(CO$_2$Me)$_2$/BSA	96	93–99 (S)	[38–40]
6	OAc	CH$_2$(CO$_2$Me)$_2$/BSA	99	97.8–98.6 (S)	[39, 40]
7	OCO$_2$Et	CH$_2$(CO$_2$Me)$_2$/BSA	–	93.8 (S)	[41]
8	OCO$_2$Et	BnNH$_2$	–	94.5 (S)	[41]
9	OAc	CH$_2$(CO$_2$Me)$_2$/BSA	>90	90–92 (S)	[42]
10	OAc	BnNH$_2$	90	99 (S)	[43]
11	OAc	CH$_2$(CO$_2$Me)$_2$/BSA	99	91 (R)	[44]
12	OAc	TsNHK or BzNHNHK	78–98	94–97 (R)	[45]
13	OAc	CH$_2$(CO$_2$Me)$_2$/BSA	97–99	91–99 (R)	[46]
14	OAc	CH$_2$(CO$_2$Me)$_2$/BSA	90–97	94–99 (S)	[47]
15	OPiv	CH$_2$(CO$_2$Me)$_2$/BSA	88–99	91–95 (R)	[48]
16 or 17	OAc	CH$_2$(CO$_2$Me)$_2$/BSA	>99	90–96 (S)	[49, 50]
18	OAc	Various amines	57–99	89–98 (S)	[51]
18	OAc	B(OMe)$_3$	92	94 (R)	[51]

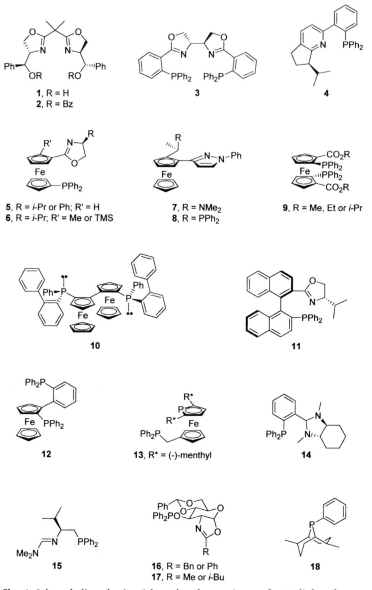

Fig. 1 Selected ligands in Pd-catalyzed reactions of 1,3-diphenylpropenyl esters or carbonates

lowed the preparation of the methyl ether in excellent yield and enantioselectivity [51]. Ligand **11** also affords excellent enantioselectivities (92.3—99% ee) for the allylic amination with various amines (benzylamine, o-MeO-C$_6$H$_4$CH$_2$NH$_2$, p-MeO-C$_6$H$_4$CH$_2$NH$_2$, and p-CH$_3$-C$_6$H$_4$SO$_2$NH$_2$/NaH) [44], whereas excellent results were also obtained using benzylamine and ligands **14** [47] or **17** [50].

Other ligands giving >90% ee for the alkylation or amination of **19** have been reported but will not be described in detail. They include derivatives of **3** [36], ferrocene-based ligands [54–57], 1,1′-binaphthyl-based ligands [58–60], natural product-based ligands such as fenchone [61], cholesterol [62], or carbohydrates [63, 64], chiral aryl chromium complexes [65, 66], chiral sulfimides [67], new P,N-ligands [19, 22, 52, 53, 68–78], and others [79–82].

The first use of non-biaryl atropisomers as chiral ligands in metal-catalyzed reactions was reported with ligand **21** [83] (Fig. 2). The allylic alkylation of **19** gave **20** in good enantioselectivity (90% ee) although in moderate yield (60%). Comparable results (89–99% yield, 90.2–94.7% ee) were obtained with **22** which possesses only axial chirality [84].

Most of the ligands reported bear phosphorous, nitrogen, or oxygen as the chelating atom. However, some ligands containing other elements including sulfur or selenium atoms in conjuncture with "typical" chelating atoms have been reported recently (Fig. 3) and showed excellent potential for inducing asymmetry, for example, in Pd-catalyzed reactions of 1,3-diphenylpropenyl acetates with dimethyl malonate (Table 2).

Some of these ligands have also been tested for the amination of 1,3-diphenylprop-2-enyl acetate. For example, ligand **29** catalyzes the addition of potassium phthalimide with good enantioselectivity (90% ee) and moderate yield (74%) [91, 92]. Ligands **24** catalyze the amination with either benzylamine or potassium phthalimide with enantioselectivities of 91–99% ee and with good yields [87]. Ligands **31** [94, 95] and **25** [88] catalyze the amination with ben-

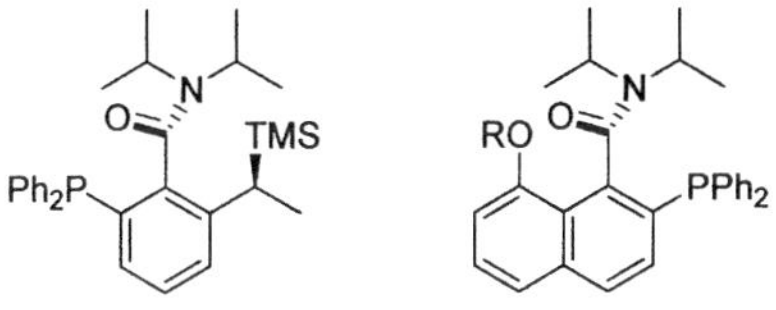

21   **22, R = Me or Bn**

Fig. 2 Non-biaryl atropisomer-based chiral ligands

Table 2

Ligand	% Yield	% ee (abs config)	Ref.
23	98–99	91–97 (R)	[85, 86]
24	90–92	96–97 (R)	[87]
25	98	98 (S)	[88]
26	71–80	98.3–99.3 (S)	[89]
27	88–96	99 (S)	[90]
28	85–90	92–99 (R)	[90]
29	80	94 (R)	[91, 92]
30	88	96 (R)	[93]
31	97	98 (S)	[94, 95]

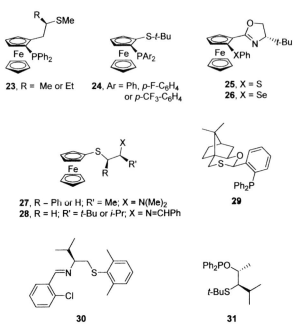

Fig. 3 List of sulfur- or selenium-containing ligands in Pd-catalyzed reactions of 1,3-diphenylpropenyl acetates with dimethyl malonate (BSA/CH$_2$(CO$_2$Me)$_2$)

zylamine with excellent results, whereas **23** (R=Et) gives 94% ee but in low yield (50%) [85, 86].

The excellent control observed with the P,S- and certain N,S-ligands is believed to be mainly electronic in origin [85–88, 90–92, 94, 95]. When a π-allyl system possesses two different coordinating atoms, the nucleophilic attack is expected to occur *trans* to the better π-acceptor, since the electronic density of the allylic system is lowest at this position. In these cases, the phosphorus and the nitrogen atoms are better π-acceptors than the sulfur which is a good donor but weak acceptor. In some cases, the sulfur atom is considered to be the better acceptor [90]. In addition, it has been proposed that the selectivity arises from subtle steric interactions that predispose attack on the allyl unit of the reaction intermediate with a preferred reaction trajectory [93].

Other nucleophiles have been used in this reaction. For example, BINAP was shown to be the best ligand for the enantioselective allylic amination of **19** with sodium diformylamide, a protected ammonia equivalent, furnishing the desired product in good yield and excellent enantioselectivity [96]. Also, a P,N-ligand has been developed for the allylic sulfonylation with good ee but moderate yield [97].

The enantioselective addition of ketene silyl acetal to **19** (X=OPiv) was shown to be possible with ligand **15** and the desired products were obtained in moderate to good yields with high enantioselectivities (90–93% ee) [48]. Alkylation of **19** using the nonstabilized pre-formed magnesium enolate of cyclohexanone (Eq. 1) and (*R*)-BINAP as the ligand gave **32** with a high degree of diastereoselectivity (98% de) and enantioselectivity (99% ee) [98]. Similarly, enantioselective and diastereoselective synthesis of unnatural protected amino acids by the addition of zinc enolate to **19** was also reported [99].

$$(1)$$

A water-soluble version of ligand **17** (R=Me) in which the phenyl groups on the phosphorus atom have been replaced by 4-(*N,N*-diethylaminomethyl)phenyl has been reported thereby allowing the recycling of the ligand at the end of the reaction by a simple liquid–liquid extraction [100]. A recyclable amphiphilic resin-supported P,N-chelating palladium complex **33** was reported to affect the asymmetric alkylation of linear and cyclic substrates with excellent results (50–94% yields, 90–98% ee) in water. The catalyst can be easily recovered at the end of the reaction by simple filtration and reused without lost of activity [101]. Asymmetric allylic alkylations have also been realized using a fluorinated derivative of MOP (2-methoxy-2′-diphenylphosphine-BINOL) in fluorinated solvent (benzotrifluoride) and toluene with good results [102]. Alkylations have been conducted in an ionic liquid with promising results [103]. Palladium complexes with a chiral P,N-ligand supported on either reversed-phase silica or mesoporous silica were shown to catalyze the allylic alkylation [104] and the amination [105], respectively. The use of a polymer-supported catalyst has been shown to be possible [106, 107]. The field of recoverable catalysts in asymmetric synthesis has been recently reviewed [108]. Finally, the asymmetric microwave-assisted Pd-catalyzed alkylation with dimethyl malonate/BSA using phosphinooxazolidine ligands has been shown to accelerate the reaction and the desired products can be obtained in excellent enantioselectivities (93% ee) and yields (65–99%) in less than 2 min [109, 110, 111].

Although less attention has been given to the kinetic resolution of **19**, interesting results have also been obtained using helical diphosphine **34** [112].

Ph₂P̈ P̈Ph₂

34

The first enantioselective alkylation of 1,3-diphenylprop-2-enyl ethylcarbonate **19** with sodio dimethyl malonate using a ruthenium-based catalyst has been reported [113]. As shown in Eq. 2, a planar-chiral cyclopentadienylruthenium complex **35** catalyses the addition in excellent yield and enantioselectivity. The complex also catalyzed the amination although in lower ee (74%). Other derivatives of the complex also gave excellent results. Interesting selectivity (90–95% ee) has been obtained with platinum complexes though the conversions are low (25–39%) [114, 115].

$$(2)$$

3.1.2
Cyclic Substrates

As mentioned earlier, although similar structurally, it is generally hard to extrapolate from the reaction of acyclic compounds to cyclic variants because of the very different steric interactions between the substrate and the catalyst. Selected ligands for the allylic alkylation or amination of **36** are shown in Fig. 4 and representative data are presented in Table 3.

Investigations by Evans revealed the complexity of extending the results from acyclic to cyclic system in the case of his P,S-ligands [95]. After optimization studies, it was found that ligand **36** was the best for the cyclic substrate, whereas **31** was optimal for acyclic compounds. The alkylation with the dimethyl malonate/BSA system or benzylamine afforded the desired products in the cycloalkenyl series (for rings with n=0–2) in high yields (>91%) and excellent enantioselectivities (>91% ee). They also investigated the alkylation of heterocyclic substrates and with **36** as the ligand showed that excellent yields and enantioselectivities were obtained when Y=NBoc for both dimethyl malonate (94%, 94% ee) and benzylamine (95%, 95% ee). When a sulfur-containing sub-

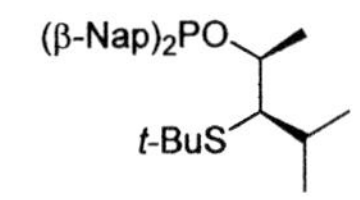

Fig. 4 Selected ligands for the Pd-catalyzed reactions of cyclic substrates

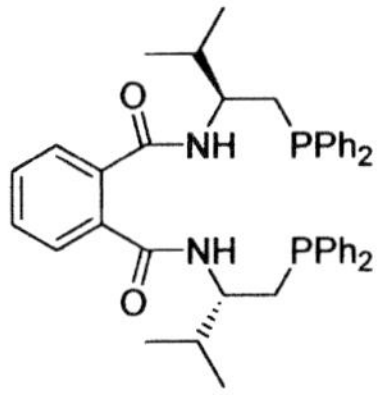

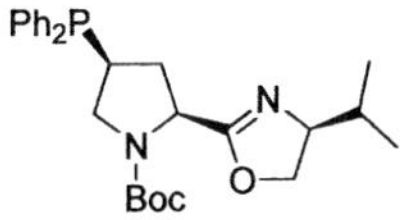

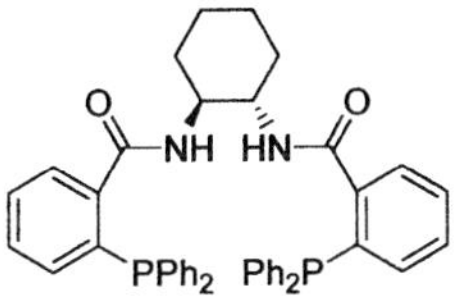

Table 3

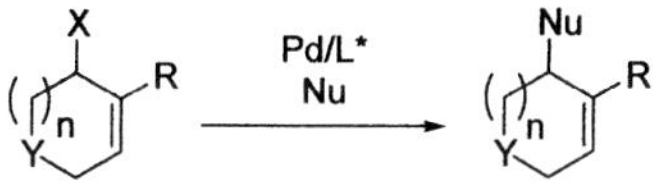

L*	X	R	Y	n	Nu	% Yield	% ee	Ref.
36	OAc	H	CH$_2$	0–2	CH$_2$(CO$_2$Me)$_2$/BSA	91–98	94–96	[95]
36	OAc	H	CH$_2$	0–2	BnNH$_2$	93–97	91–97	[95]
36	OCO$_2$Et	H	N-Boc	1	BnNH$_2$	95	94	[95]
37	OPiv	H	CH$_2$	1	CH$_2$(CO$_2$Me)$_2$/BSA	64	>99	[48, 116]
38	OAc	H	CH$_2$	0	CH$_2$(CO$_2$Me)$_2$/BSA	79–99	90–96	[117, 118]
18	OCO$_2$Me	Ph	CH$_2$	1	CH$_2$(CO$_2$Me)$_2$/BSA	96–97	90–95	[119]
18	OCO$_2$Me	Ph	CH$_2$	1	TsNHCH$_2$CO$_2$Et	77	97	[119]
4	OPiv	H	CH$_2$	2	CH$_2$(CO$_2$Me)$_2$/BSA	92–98	91–94	[120]
39	OCO$_2$Me	H	CH$_2$	0–2	CH$_3$NO$_2$/BSA	94–99	95–99	[121]

strate was used (Y=S), the alkylation with dimethyl malonate gave only modest yield and selectivity (60%, 50% ee), whereas the amine proceeded with excellent enantioselectivity (92% ee) but modest yield (62%). Alkylation of **40** (R=H, Y=CH$_2$, n=0) with dimethyl malonate and ligand **39** afforded the desired product in excellent yield and enantioselectivity (94–99%, 90–96% ee) but only moderate selectivity ($\approx$80% ee) was observed for larger rings (6- and 7-membered rings) [117, 118]. Trost has shown that nitromethane can be efficiently used as a nucleophile in conjunction with ligand **39**. Other nitroalkanes also add to **40** in good yields and excellent enantioselectivities [121]. Interesting results for the kinetic resolution of this class of substrate have also been obtained [122].

A ruthenium-based catalyst has been shown to catalyze the alkylation of **40** with soft nucleophiles [123]. Addition of hard nucleophiles to cyclic substrates has also been studied using nickel catalysts. Although limited success was achieved with arylboronic acids [124], interesting selectivity (91–95% ee) has been obtained with Grignard reagents using optically active oxazolinylferrocenylphosphines although the yields are moderate (50–73%) [125].

The alkylation of derivatives of **40** have been used as the key step for total syntheses of (-)-dehydrotubifoline and (-)-tubifoline [126], the first enantioselective synthesis of (-)-wine lactone [127], and an indolizidine alkaloid [128]. Piperidine derivatives have also been obtained through this strategy [129]. Cyclopentobarbital, a pharmaceutical, has been synthesized using barbituric acid derivatives as the nucleophile [130]. The first- and second-generation asymmetric syntheses of (-)-cyclophellitol were realized [131]. Finally, (-)-galanthamine (**46**) has been successfully synthesized using this strategy (Scheme 2) [132]. In this case, the cyclic substrate **43** is alkylated by the phenol **42** under palladium catalysis with ligand **45** to furnish the desired product **44** in good yield and enantioselectivity.

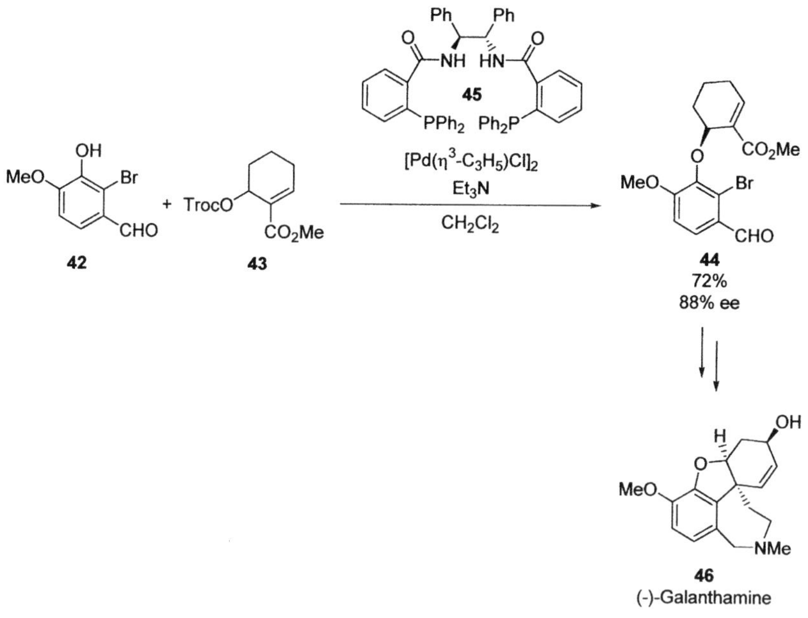

Scheme 2

3.2
Substrates with Different Substituents at C1 and C3 (R^{1}CH=CH–CHR^{2}X)

For substrates bearing different substituents at C1 and C3, the conversion of a racemic mixture to a single enantiomerically enriched product is usually not possible [1]. However, chiral catalysts have been used in this case to control the regioselectivity in reactions with non-racemic substrates, as illustrated in Scheme 3 [133]. Starting from (R)-47 either (S)-48 or the regioisomer (R)-49 can be selectively prepared, depending on the chirality of the ligand. It is not necessary to start from an enantiomerically pure substrate because the major and minor enantiomers are converted to different regioisomers (not to enantiomeric products), resulting in products of high ee even if the starting material is only of moderate enantiomeric purity. The influence of the chiral ligand clearly dominates over steric and electronic effects of the allyl substituents, as seen from the preferred formation of (R)-49 with the (R)-phosphinooxazoline ligand, in contrast to the reaction with an achiral ligand such as triphenylphosphine, which favors the other regioisomer (S)-48. Thus, the use of chiral non-racemic substrates in combination with a chiral enantiopure catalyst is a powerful strategy for controlling the regioselectivity in allylic substitutions of this type.

The synthesis of γ-fluoroalkylated allylic alcohols and amines like 51 starting with chiral fluorinated allylic mesylates 50 has also been reported (Eq. 3) [134]. In this case, the regiochemistry of the addition is controlled by the substrate and the addition of the nucleophile occurs distal to the fluorinated alkyl chain.

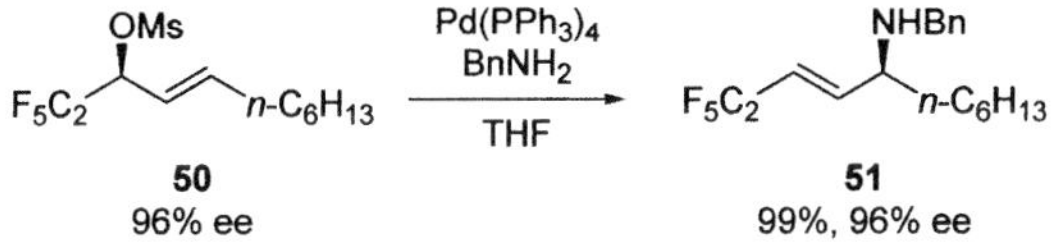

$$\text{(3)}$$

The formation of amino acid derivatives by addition of a zinc enolate to non-racemic unsymmetrical substrates has also been described, in which π–σ–π

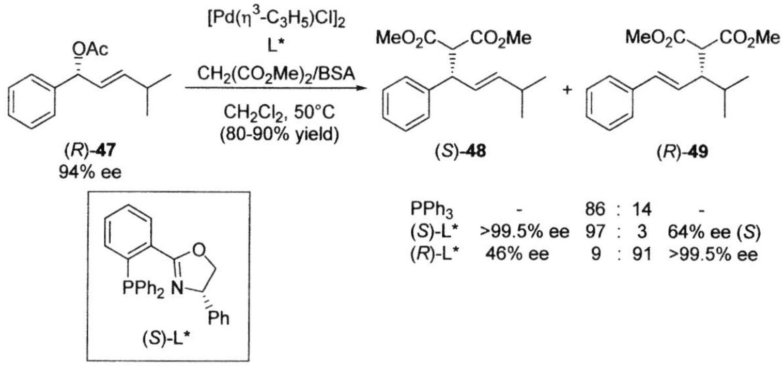

PPh$_3$	-	86 : 14	-
(S)-L*	>99.5% ee	97 : 3	64% ee (S)
(R)-L*	46% ee	9 : 91	>99.5% ee

Scheme 3

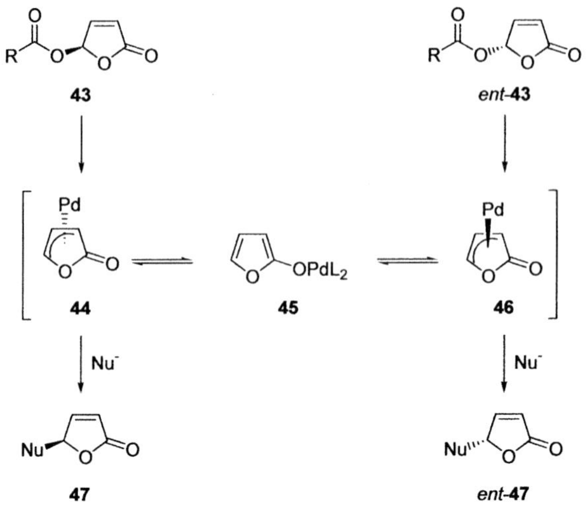

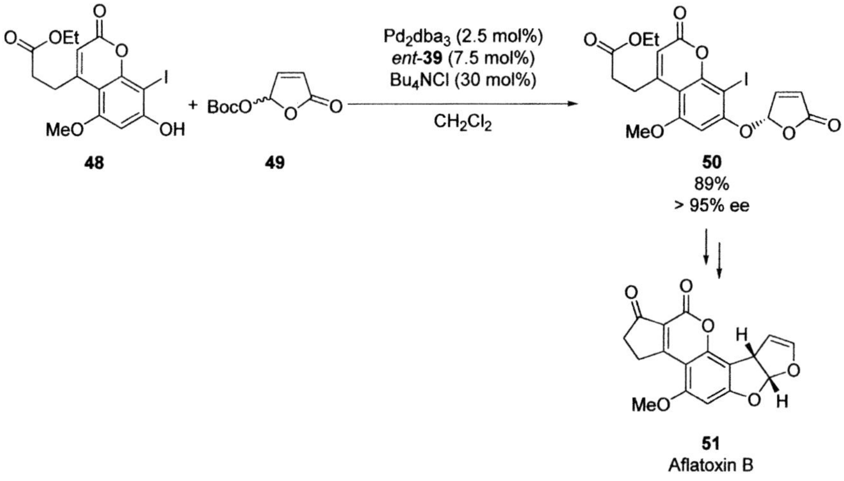

Scheme 4

isomerization is suppressed at low temperature (-78°C) and attack of the nucleophile occurs at the least hindered position [135, 136].

A special case where a racemic mixture can be converted to a single enantiomer was reported by Trost [137]. In this instance, the interconversion of the π-allyl complexes **44** and **45** is possible through a furan intermediate and the configuration of the product is controlled by the chiral ligand (Scheme 4). This strategy was used for the formal synthesis of (-)-aflatoxin B lactone (**51**) where the addition product **50** is obtained by the reaction of phenol **48** with furanone **49** under palladium catalysis with ligand *ent*-**39**.

3.3
Substrates with Identical Geminal Substituents at C1 or C3 ($R^1CH=CH-C(R^2)_2X$) or $(R^1)_2C=CH-CHR^2X$

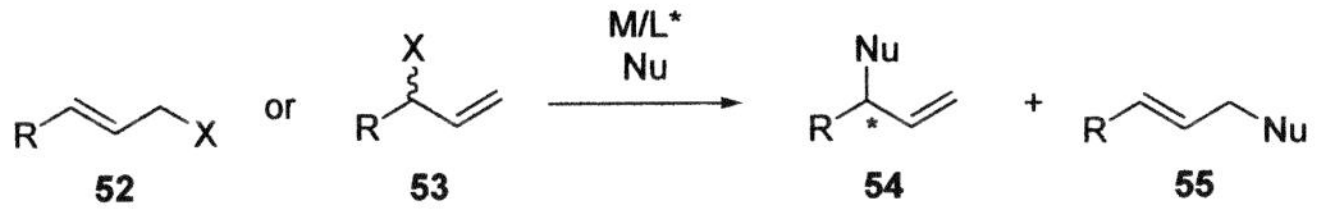

(4)

Bis(*N*-tosylamino)phosphines **56** promote the alkylation of **52** (R=Ph or 1-naphthyl, X=OAc) with dimethyl malonate/BSA to yield the desired branched product as the major component (**54/55**=60:40 to 98:2) in good yield (85–95%) and excellent enantioselectivity (94–99.4% ee) (Fig. 5) [138]. The ferrocene-based ligand **57** has also been successfully applied to the alkylation of **52** [139]. In this case, a variety of substrates (**52**, R=phenyl, 1-naphthyl, 4-substituted aryl, X=OAc) gave excellent yields (91–98%), regioselectivities (**54/55**=90:10 to >99:1), and enantioselectivities (92–97% ee). Interestingly, an excellent result was obtained with R=Me where the branched product was obtained almost as the sole product in 83% yield and 94% ee. Allylic amination of **53** with benzylamine was also possible with a variety of aryl substituents using a slightly modified ligand. The branched products **54** were obtained in good yields (76–94%) and excellent enantioselectivities (94–98% ee) in preference for the linear products **55** (monoalkylated and dialkylated). The sub-class $(R^1)_2C=CH-CHR^2X$ has also been studied in the context where R^1=Ph and good results where obtained with a derivative of **36** for R^2=Ph [95] and **58** for various R groups [140].

The conversion of racemic butadiene monoepoxide to a single enantiomer of **54** (R=CH$_2$OH) using Pd-catalyzed asymmetric allylic alkylation was uncovered by Trost [141]. By using a slightly modified ligand, the alkylation with phthalim-

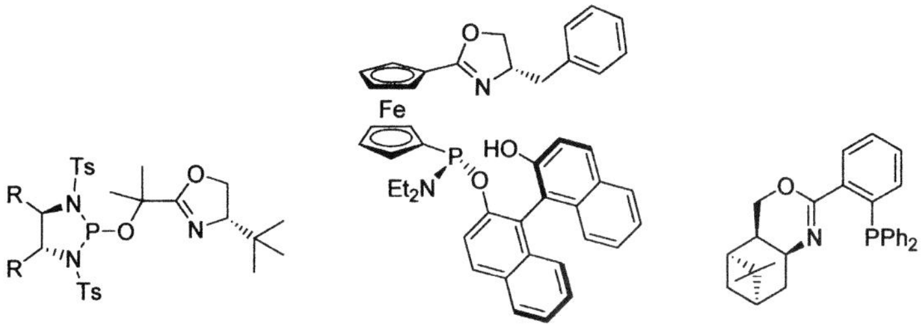

56, R = Ph or (CH$_2$)$_4$ **57**   **58**

Fig. 5

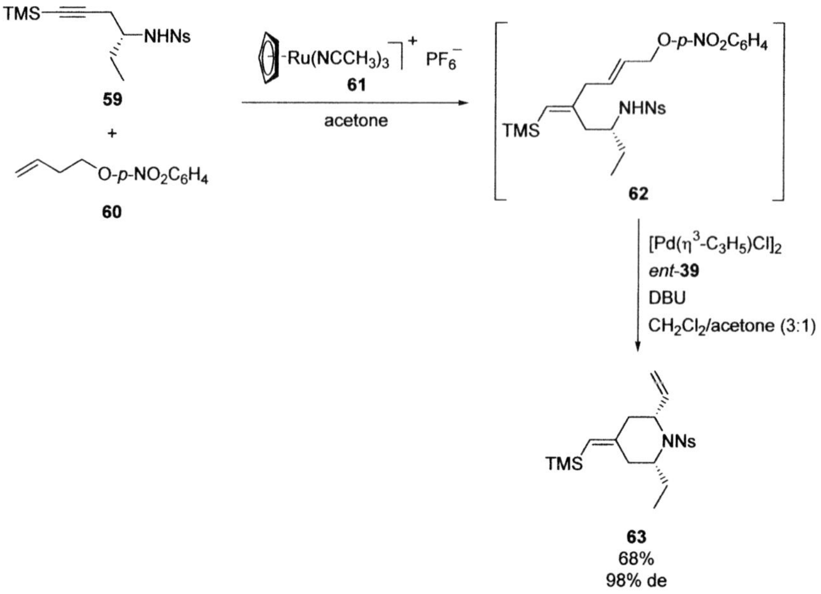

Scheme 5

ide allowed the synthesis of vigabatrin, an anti-epileptic drug, and ethambutol, a tuberculostatic drug. By using a similar strategy, they were able to accomplish the total synthesis of (-)-malyngolide [142]. Inorganic carbonates have been shown to add to isoprene monoepoxide to afford vinylglycidols in good yields and enantioselectivity [143]. An analogous approach has been applied to the deracemization of Baylis–Hillman adducts with reasonable results [144]. Recently, Trost has reported a one-pot enantioselective and diastereoselective synthesis of heterocycles by a Ru-catalyzed ene-yne coupling followed by a Pd-catalyzed asymmetric allylic alkylation sequence [145]. An example is shown in Scheme 5, where alkyne **59** is first reacted with **60** using a ruthenium catalyst (**61**) to generate **62** by an ene-yne coupling. The intermediate **62** is not isolated and is submitted to the allylic alkylation protocol using **39** as the ligand to generate the desired piperidine derivative **63** in reasonable yield and excellent enantioselectivity and diastereoselectivity. A derivative of catalyst **61** (η^5-C$_5$Me$_5$ instead of η^5-C$_5$H$_5$) also promotes allylic alkylation where the stereochemical outcome is dictated by the configuration of the starting carbonate, since the substitution occurs with net retention [146].

Total syntheses of callipeltoside A [147] and the vitamin E core [148] have been published in which one of the key steps is an asymmetric allylic alkylation of this class of substrate using a phenol derivative as nucleophile.

Other metals have also been recently used with this class of substrate. For example, a molybdenum catalyst and ligand **64** promote the alkylation between **52** (R=Ph, X=OCO$_2$Me) and dimethyl malonate to provide the branched prod-

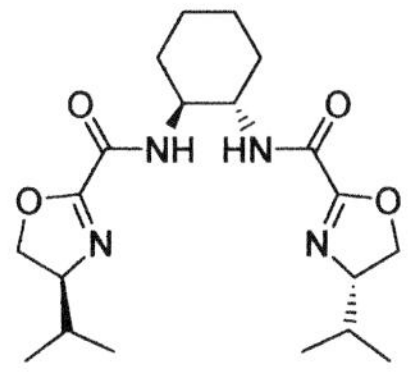

Fig. 6

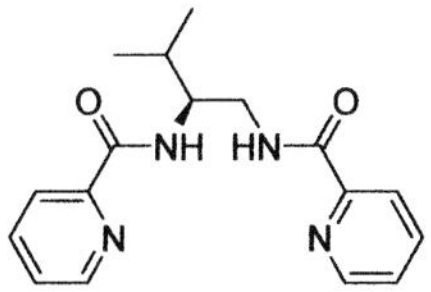

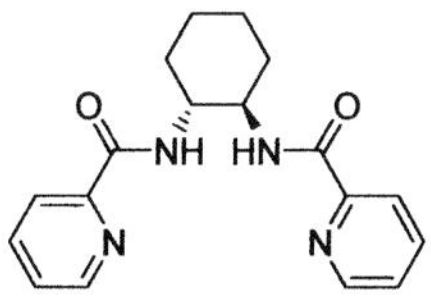

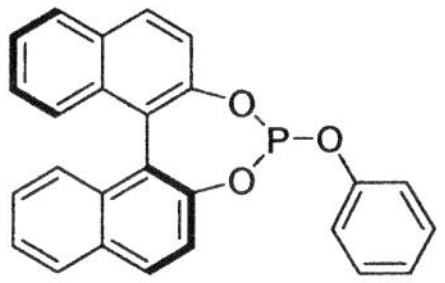

Fig. 7

uct (14:1 over the linear) with 86% yield and 99% ee (Fig. 6) [149, 150]. Excellent results (>80% yield, 92–97% ee) were also obtained with **52** (R=Me) [150]. For other substrates (**52**, R=Me, X=OCO$_2$Me, Nu=dimethyl malonate or **52**, R=OPh, X=OAc, Nu=methyl dimethyl malonate) a derivative of ligand **64** was used, giving the branched product (>9:1) in good yield (78–81%) and excellent enantioselectivities (95–98% ee). Ligand **65** was also shown to afford the branched product in moderate to good yields (59–87%) and excellent enantioselectivities (92–98% ee) for the alkylation with dimethyl malonate of **52** (R=Ph, X=OCO$_2$Me) [151]. The Mo-catalyzed alkylations have been extensively studied including the effect of substitution of the Trost ligand (**66**, substituted pyridines) [152, 153], the precatalyst used [154], the use of microwaves [155], and the role of CO transfer in the catalytic cycle [156]. Kinetic studies with the Trost catalyst in different solvents revealed a significant stereochemical memory effect in the reaction of **53** with dimethyl malonate, resulting in lower ee with increasing conversion. This implies that equilibration of the intermediate oxidative addition complexes and the subsequent nucleophilic addition step have similar rates [157]. However, in toluene and dichloromethane, where memory effects are small, high ee and full conversion could be achieved.

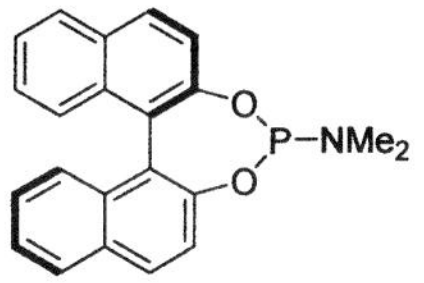

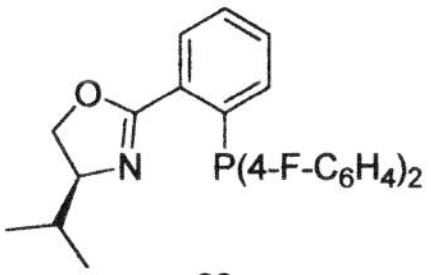

Iridium has also been used in conjunction with ligand **67** for enantioselective alkylation reactions (Fig. 7) [158]. For example, **52** (R=Ph, X=OCO$_2$Me) reacts with dimethyl malonate to give the branched product in 99% yield and 96% ee. The enantiomeric excess was highly dependent on the base used, since only BuLi/ZnCl$_2$ gave enantiomeric excesses >36%. Other ligands, including **68** and **69**, also gave good results [159, 160]

Allylic amination of **70** using iridium catalyst has also been reported recently using a chiral phophoramidite ligand. The reaction produced the branched product **72** over the linear monoalkylated and dialkylated products and the amines

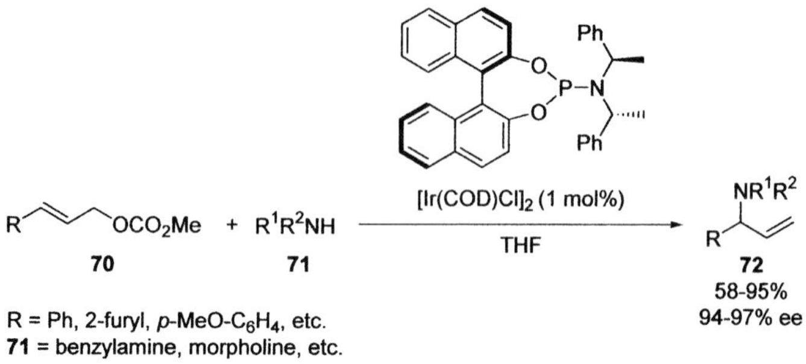

R = Ph, 2-furyl, *p*-MeO-C$_6$H$_4$, etc.
71 = benzylamine, morpholine, etc.

Scheme 6

were isolated in good yields and excellent enantiomeric excesses (Scheme 6) [161].

The synthesis of monosubstituted and disubstituted pyrrolidines has been shown to be possible using a regioselective Rh-catalyzed allylic amination/ring-closing metathesis sequence [162, 163] starting from enantiopure **54**. Tandem Rh-catalyzed allylic amination/Pauson–Khand annulation reactions have also been studied [164]. Finally, the synthesis of *anti*-1,3-carbon stereogenic centers and C_2-symmetric fragments using a Rh-catalyzed allylic linchpin cross-coupling reaction was shown to be possible. For example, compound **75** can be obtained from the alkylation of carbonate **74** with the malonate derivative **73**, as shown in Eq. 5 [165].

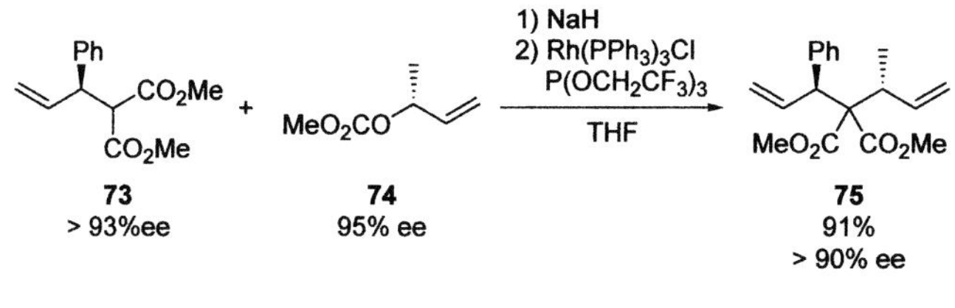

(5)

3.4
meso-Substrates with Two Enantiotopic Leaving Groups

Evans showed that his P,S-ligand **36** was also efficient for the monoalkylation of *cis*-1,4-diacetoxycyclopentene **76**. Reaction of dimethyl malonate/BSA gave **77** in 85% yield and 96% ee (Eq. 6) [95]. The same reaction and the intramolecular cyclization of a biscarbamate has also been reported using a polymer-supported Trost-type catalyst with excellent results [166, 167].

$$\text{76} \xrightarrow[\substack{\text{CH}_2\text{Cl}_2, \, -20°\text{C}}]{\substack{[\text{Pd}(\eta^3\text{-C}_3\text{H}_5)\text{Cl}]_2 \\ \textbf{36} \\ (\text{CH}(\text{CO}_2\text{Me})_2/\text{BSA}}} \text{77}$$

AcO⟨⟩OAc
76

AcO⟨⟩CO_2Me / CO_2Me
77
85%
96% ee

(6)

The enantioselective syntheses of carbanucleosides (e.g., **80**) have also been reported where the key step is the enantioselective allylic amination of **78** with a nucleobase (here **79**) (Eq. 7) [168]. A chiral phosphine bearing a carboxyl group has also been shown to be effective for this class of substrate [169].

BocO⟨⟩OBoc + [purine **79**] $\xrightarrow[\text{DMF}]{\substack{\text{Pd}_2(\text{dba})_3\cdot\text{CHCl}_3 (1 \text{ mol\%}) \\ \textbf{45} \, (3 \text{ mol\%})}}$ BocO⟨⟩**80**

78   **79**   **80**
                                                                     62%
                                                                     > 98% ee

(7)

The synthesis of C-2-*epi*-hydromycin A [170] and tetraponerines [171] using a desymmetrization of **76** (bearing benzoates and carbonates respectively instead of acetate) has been reported. Finally, the first-generation and second-generation asymmetric syntheses of the aminocyclohexitol moiety of hygromycin A were reported [172].

Table 4

$$\text{81} \xrightarrow[\text{THF}]{\substack{\text{R} \diagup \text{EWG} / \text{EWG} \\ [\text{Pd}(\eta^3\text{-C}_3\text{H}_5)\text{Cl}]_2 \\ \textbf{39}}} \text{82}$$

TBDPSO⟍⟍OAc / OAc
81

TBDPSO⟍⟍ OAc / EWG / EWG / R
82

Nu		% Yield	% ee
R	EWG		
CH$_3$	CO$_2$Me	87	93
CH$_2$C≡CH	CO$_2$Me	86	91
OMOM	CO$_2$Me	79	93
NHTroc	CO$_2$Me	92	89
CH$_2$Ph	CN	80	90
CH$_3$	SO$_2$Ph	58	92

3.5 Substrates with Two Geminal Enantiotopic Leaving Groups

The allylic alkylation of geminal leaving groups has been recently studied by Trost. He showed that the reaction was highly influenced by the reaction conditions including nucleophile counter-ion, ligand, leaving group, solvent, concentration, and temperature [173, 174]. Under the optimized conditions, the reaction is quite general and a few examples of nucleophiles are shown in Table 4. This process constitutes the equivalent of an addition of a stabilized nucleophile to a carbonyl group with high asymmetric induction. They also showed that the product could be used in subsequent transformations including allylic transposition, a second allylic alkylation with malonate derivatives and phthalimide, and Claisen rearrangement giving access to useful chiral building blocks.

3.6 Reactions of Prochiral Nucleophiles

Recently, Trost reported an efficient Pd-catalyzed asymmetric allylic alkylation of 1-tetralones [175] and α-arylketone **83** creating a quaternary center as exemplified by the synthesis of **84** (Scheme 7) [176]. Two α-heterocyclic ketones were also alkylated with similar results. Ferrocene-based ligands are effective in pro-

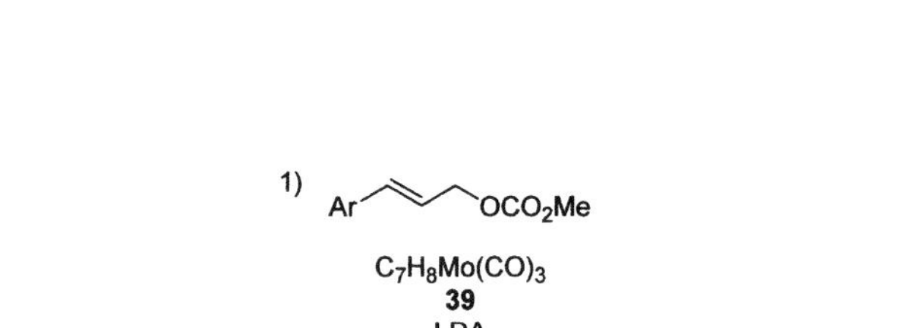

O
Ar
+
OAc
NaHMDS
[Pd(η3-C3H5)Cl]2
39
DME, 0°C
n
83, Ar = Ph, p-MeO-C6H4, etc.
n = 1-3
84
77-95%
83-92% ee

Scheme 7

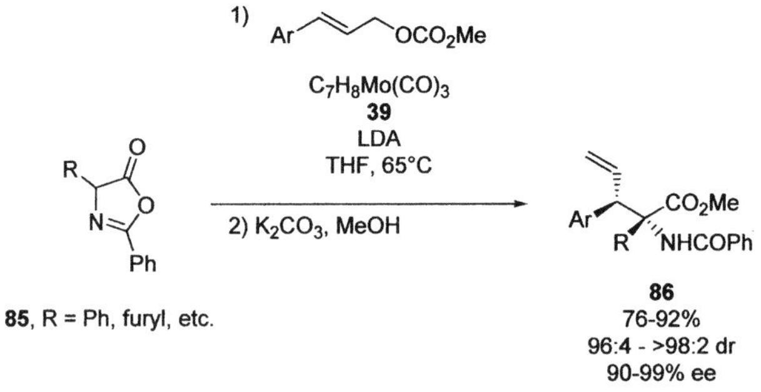

1)
Ar
OCO2Me
C7H8Mo(CO)3
39
LDA
THF, 65°C
2) K2CO3, MeOH
R
O
N
Ph
CO2Me
Ar
R
NHCOPh
85, R = Ph, furyl, etc.
86
76-92%
96:4 - >98:2 dr
90-99% ee

Scheme 8

moting this reaction [177]. The alkylation of β-ketoesters to afford quaternary centers has also been reported [178].

The synthesis of quaternary amino acids **86** have been shown using azlactones **85** as nucleophiles and the Trost ligand **39** under palladium [179] or molybdenum catalysis (Scheme 8) [180]. The allylic alkylation of glycine imino esters under biphasic conditions has also been achieved using a chiral phase-transfer catalyst in combination with an achiral Pd catalyst producing the unnatural amino acid derivatives [181].

4
Conclusion

An enormous amount of work has been done in the field of allylic substitution reactions in the past four years as exemplified by this update. Good results can be obtained in almost every class of substrates; however, the scope of nucleophiles and substrates used is still relatively narrow. Unsymmetrical substrates, substrates with geminal leaving groups, and reactions producing two controlled stereocenters are largely unexplored and deserve more attention, since the products generated are of greater interest. Clearly, the allylic alkylation reaction is one of the most powerful reactions for the construction of C–C and C–heteroatom bonds and future research should aim at understanding and enhancing its potential.

References

1. Pfaltz A, Lautens M (1999) Comprehensive asymmetric catalysis. In: Jacobsen EN, Pfaltz A, Yamamoto H (eds) Springer, Berlin Heidelberg New York, Vol 2, Chap 24
2. Tsuji J (1999) Perspective in organopalladium chemistry for the XXI century. Elsevier, Amsterdam
3. Trost BM, Chulbom L (2000) In: Ojima I (ed) Catalytic asymmetric synthesis. Wiley, New York, Chap 8E
4. Tsuji J (2000) Transition metal reagents and catalysts. Innovation in organic synthesis. Wiley, New York, Chap 4
5. Negishi E-I (2002) Handbook of organopalladium chemistry for organic synthesis. Wiley, New York
6. Trost BM (2002) Chem Pharm Bull 50:1
7. Hayashi T (1999) J Organomet Chem 576:195
8. Helmchen G (1999) J Organomet Chem 576:203
9. Koovsk P, Malkov AV, Vyskoil , Lloyd-Jones GC (1999) Pure Appl Chem 71:1425
10. Moberg C, Bremberg U, Hallman K, Svensson M, Norrby P-O, Hallberg A, Larhed M, Csöregh I (1999) Pure Appl Chem 71:1477
11. Pfaltz A (2001) Chimia 55:708
12. Fache F, Shulz E, Tammasino ML, Lemaire M (2000) Chem Rev 100:2159
13. Helmchen G, Pfaltz A (2000) Acc Chem Res 33:336
14. Hayashi T (2000) Acc Chem Res 33:354
15. Drago D, Pregosin PS (2000) J Chem Soc Dalton Trans 3191
16. Canal JM, Gómez M, Jiménez F, Rocamora M, Muller G, Duñach E, Franco D, Jiménez A, Cano FH (2000) Organometallics 19:966

17. Kawatsura M, Uozumi Y, Ogasawara M, Hayashi T (2000) Tetrahedron 56:2247
18. Junker J, Reif B, Steinhagen H, Junker B, Felli IC, Reggelin M, Griesinger C (2000) Chem Eur J 6:3281
19. Liu S, Müller JFK, Neuburger M, Schaffner S, Zehnder M (2000) Helv Chim Acta 83:1256
20. Ogasawara M, Takizawa K-I, Hayashi T (2002) Organometallics 21:4853
21. Kollmar M, Steinhagen H, Janssen JP, Goldfuss B, Malinovskaya SA, Vázquez J, Rominger F, Helmchen G (2002) Chem Eur J 8:3103
22. Zehnder M, Schaffner S, Neuburger M, Plattner DA (2002) Inorg Chim Acta 337:287
23. Widhalm M, Nettekoven U, Kalchhauser H, Mereiter K, Calhorda MJ, Félix V (2002) Organometallics 21:315
24. Dotta P, Kumar PGA, Pregosin PS (2002) Magn Reson Chem 4:653
25. Svensson M, Bremberg U, Hallman K, Csöregh I, Moberg C (1999) Organometallics 18:4900
26. Trost BM, Toste FD (1999) J Am Chem Soc 121:4545
27. Lloyd-Jones GC, Stephen SC, Murray M, Butts CP, Vyskoil , Koovsk P (2000) Chem Eur J 6:4348
28. Fairlamb IJS, Lloyd-Jones GC, Vyskoil , Koovsk P (2002) Chem Eur J 8:4443
29. Fairlamb IJS, Lloyd-Jones GC (2000) Chem Commun 2447
30. Goldfuss B, Kazmaier U (2000) Tetrahedron 56:6493
31. Amatore C, Jutand A, M'Barki MA, Meyer G, Mottier L (2001) Eur J Inorg Chem 873
32. Solin N, Szabó KJ (2001) Organometallics 20:5464
33. Tsurugi K, Nomura N, Aoi K (2002) Tetrahedron Lett 43:469
34. Hoarau O, Aït-Haddou H, Daran J-C, Cramailère D, Balavoine GGA (1999) Organometallics 18:4718
35. Lee S-G, Lim CW, Song CE, Kim KM, Jun CH (1999) J Org Chem 64:4445
36. Lee S-G, Lee SH, Song CE, Chung BY (1999) Tetrahedron Asymmetry 10:1795
37. Ito K, Kashiwagi R, Iwasaki K, Katsuki T (1999) Synlett 1563
38. Park J, Quan Z, Lee S, Ahn KH, Cho C-W (1999) J Organomet Chem 584:140
39. Deng W-P, You S-L, Hou X-L, Dai L-X, Yu Y-, Xia W, Sun J (2001) J Am Chem Soc 123:6508
40. Deng W-P, Hou X-L, Dai LX, Yu Y-H, Xia W (2000) Chem Commun 285
41. Burckhard U, Drommi D, Togni A (1999) Inorg Chim Acta 296:183
42. Zhang W, Shimanuki T, Kida T, Nakatsuji Y, Ikeda I (1999) J Org Chem 64:6247
43. Nettekoven U, Widhalm M, Kalchhauser H, Kamer PCJ, van Leeuwen PWNM, Lutz M, Spek AL (2001) J Org Chem 66:759
44. Selvakumar K, Valentini M, Wörle M, Pregosin PS, Albinati A (1999) Organometallics 18:1207
45. Lotz M, Kramer G, Knochel P (2002) Chem Commun 2546
46. Ogasawara M, Yoshida K, Hayashi T (2001) 20:3913
47. Jin M-J, Kim S-H, Lee S-J, Kim Y-M (2002) Tetrahedron Lett 43:7409
48. Saitoh A, Achiwa K, Tanaka K, Morimoto T (2000) J Org Chem 65:4227
49. Yonehara K, Hashizume T, Mori K, Ohe K, Uemura S (1999) Chem Commun 415
50. Yonehara K, Hashizume T, Mori K, Ohe K, Uemura S (1999) J Org Chem 64:9374
51. Hamada Y, Seto N, Takayanagi Y, Nakano T, Hara O (1999) Tetrahedron Lett 40:7791
52. Anderson JC, Cubbon RJ, Harling JD (1999) Tetrahedron Asymmetry 10:2829
53. Anderson JC, Cubbon RJ, Harling JD (2001) Tetrahedron Asymmetry 12:923
54. Fukuda T, Takehara A, Iwao M (2001) Tetrahedron Asymmetry 12:2793
55. Kang J, Lee JH, Choi JS (2001) Tetrahedron Asymmetry 12:33
56. Lee S, Koh JH, Park H (2001) J Organomet Chem 637–639:99
57. Hu X, Dai H, Hu X, Chen H, Wang J, Bai C, Zheng Z (2002) Tetrahedron Asymmetry 13:1687
58. Kodama H, Taiji T, Ohta T, Furukawa I (2000) Tetrahedron Asymmetry 11:4009
59. Reetz MT, Haderlein G, Angermund K (2000) J Am Chem Soc 122:996
60. Wang Y, Guo H, Ding K (2000) Tetrahedron Asymmetry 11:4153
61. Suzuki Y, Ogata Y, Hiroi K (1999) Tetrahedron Asymmetry 10:1219

62. Chelucci G, Pinna GA, Saba A, Sanna G (2000) J Mol Cat A Chemical 159:423
63. Pàmies O, van Strijdonck GPF, Diéguez M, Deerenberg S, Net G, Ruiz A, Claver C, Kamer PCJ, van Leeuwen PWNM (2001) J Org Chem 66:8867
64. Diéguez M, Jansat S, Gomez M, Ruiz A, Muller G, Claver C (2001) Chem Commun 1132
65. Han JW, Jang H-Y, Chung YK (1999) Tetrahedron Asymmetry 10:2853
66. Jang H-Y, Seo H, Han JW, Chung YK (2000) Tetrahedron Lett 41:5083
67. Takada H, Oda M, Oyamada A, Ohe K, Uemura S (2000) Chirality 12:299
68. Jin M-J, Jung J-A, Kim S-H (1999) Tetrahedron Lett 40:5197
69. Mino T, Tanaka Y, Sakamoto M, Fujita T (2001) Tetrahedron Asymmetry 12:2435
70. Mino T, Ogawa T, Yamashita M (2001) Heterocycles 55:453
71. Mino T, Shiotsuki M, Yamamoto N, Suenaga T, Sakamoto M, Fujita T, Yamashita M (2001) J Org Chem 66:1795
72. Mino T, Ogawa T, Yamashita M (2003) J Organomet Chem 665:122
73. Bernardinelli GH, Kündig EP, Meier P, Pfaltz A, Radkowski K, Zimmermann N, Neuburger-Zehnder M (2001) Helv Chim Acta 84:3233
74. Kondo K, Kazuta K, Fujita H, Sakamoto Y, Murakami Y (2002) Tetrahedron 58:5209
75. Okuyama Y, Nakano H, Hongo H (2000) Tetrahedron Asymmetry 11:1193
76. Hou D-R, Reibenspies JH, Burgess K (2001) J Org Chem 66:206
77. Gómez M, Jansat S, Muller G, Panyella D, van Leeuwen PWNM, Kamer PCJ, Goubitz K, Fraanje J (1999) Organometallics 18:4970
78. Saitoh A, Misawa M, Morimoto T (1999) Synlett 483; Saitoh A, Misawa M, Morimoto T (1999) Synlett
79. Chelucci G, Medici S, Saba A (1999) Tetrahedron 10:543
80. Chelucci G, Deriu SP, Saba A, Valenti R (1999) Tetrahedron Asymmetry 10:1457
81. Chelucci G, Deriu S, Pinna GA, Saba A, Valenti R (1999) Tetrahedron Asymmetry 10:3803
82. Inoue H, Nagaoka Y, Tomioka K (2002) J Org Chem 67:5864
83. Clayden J, Johnson P, Pink JH, Helliwell M (2000) J Org Chem 65:7033
84. Dai W-M, Yeung KKY, Liu J-T, Zhang Y, Williams ID (2002) Org Lett 4:1615
85. Enders D, Peters R, Runsink J, Bats JW (1999) Org Lett 1:1863
86. Enders D, Peters R, Lochtman R, Raabe G, Runsink J, Bats JW (2000) Eur J Org Chem 3399
87. Priego J, Mancheño OG, Cabrera S, Arrayás RG, Llamas T, Carretero JC (2002) Chem Commun 2512
88. You S-L, Hou X-L, Dai L-X, Yu Y-H, Xia W (2002) J Org Chem 67:4684
89. You S-L, Hou X-L, Dai L-X (2000) Tetrahedron Asymmetry 11:1495
90. Bernardi L, Bonini BF, Comes-Franchini M, Fochi M, Mazzanti G, Ricci A, Varchi G (2002) Eur J Org Chem 2776
91. Nakano H, Okuyama Y, Hongo H (2000) Tetrahedron Lett 41:4615
92. Nakano H, Okuyama Y, Yanagida M, Hongo H (2001) J Org Chem 66:620
93. Adams H, Anderson JC, Cubbon R, James, DS, Mathias, JP (1999) J Org Chem 64:8256
94. Evans DA, Campos KR, Tedrow JS, Michael FE, Gagné MR (1999) J Org Chem 64:2994
95. Evans DA, Campos KR, Tedrow JS, Michael FE, Gagné MR (2000) J Am Chem Soc 122:7905
96. Wang Y, Ding K (2001) J Org Chem 66:3238
97. Bondarev OG, Lyubimov SE, Shiryaev AA, Kadilnikov NE, Davankov VA, Gavrilov KN (2002) Tetrahedron Asymmetry 13:1587
98. Braun M, Laicher F, Meier T (2000) Angew Chem Int Ed 39:3494
99. Kazmaier U, Zumpe FL (2000) Angew Chem Int Ed 39:802
100. Hashizume T, Yonehara K, Ohe K, Uemura S (2000) J Org Chem 65:5197
101. Uozumi Y, Shibatomi K (2001) J Am Chem Soc 123:2919
102. Cavazzini M, Pozzi G, Quici S, Maillard D, Sinou D (2001) Chem Commun 1220
103. Kmentová I, Gotov B, Solcániová E, Toma (2002) Green Chem 4:103
104. Anson MS, Mirza AR, Tonks L, Williams JMJ (1999) Tetrahedron Lett 40:7147

105. Johnson BFG, Raynor SA, Shepard DS, Mashmeyer T, Thomas JM, Sankar G, Bromley S, Oldroyd R, Gladden L, Mantle MD (1999) Chem Commun 1167
106. Saluzzo C, ter Halle R, Touchard F, Fache F, Shulz E, Lemaire M (2000) J Organomet Chem 603:30
107. Akiyama R, Kobayashi S (2001) Angew Chem Int Ed 40:3469
108. Fan Q-H, Li Y-M, Chan ASC (2002) Chem Rev 102:3385
109. Bremberg U, Lutsenko S, Kaiser N-F, Larhed M, Hallberg A, Moberg C (2000) Synthesis 1004
110. Kaiser N-FK, Bremberg U, Larhed M, Moberg C, Hallberg A (2000) J Organomet Chem 603:2
111. Bremberg U, Larhed M, Moberg C, Hallberg A (1999) J Org Chem 64:1082
112. Reetz MT, Sostmann S (2000) J Organomet Chem 603:105
113. Matsushima Y, Onitsuka K, Kondo T, Mitsudo T-A, Takahashi S (2001) J Am Chem Soc 123:10405
114. Blacker AJ, Clarke ML, Loft MS, Williams JMJ (1999) Chem Commun 913
115. Blacker AJ, Clarke ML, Loft MS, Mahon MF, Humphries ME, Williams JMJ (2000) Chem Eur J 6:353
116. Saitoh A, Misawa M, Morimoto T (1999) Tetrahedron Asymmetry 10:1025
117. Gilbertson SR, Xie D (1999) Angew Chem Int Ed 38:2750
118. Gilbertson SR, Xie D, Fu Z (2001) J Org Chem 66:7240
119. Hamada Y, Sakaguchi K-E, Hatano K, Hara O (2001) Tetrahedron Lett 42:1297
120. Ito K, Kashiwagi R, Hayashi S, Uchida T, Katsuki T (2001) Synlett 284
121. Trost BM, Surivet J-P (2000) Angew Chem Int Ed 39:3122
122. Longmire JM, Wang B, Zhang X (2000) Tetrahedron Lett 41:5435
123. Morisaki Y, Kondo T, Mitsudo T-A (1999) Organometallics 18:4742
124. Chung K-G, Miyake Y, Uemura S (2000) J Chem Soc Perkin Trans 1 15
125. Chung K-G, Miyake Y, Uemura S (2000) J Chem Soc Perkin Trans 1 2725
126. Mori M, Nakanishi M, Kajishima D, Sato Y (2001) Org Lett 3:1913
127. Bergner EJ, Helmchen G (2000) Eur J Org Chem 419
128. Ovaa H, Stragies R, van der Marel GA, van Boom JH, Blechert S (2000) Chem Commun 1501
129. Schleich S, Helmchen G (1999) Eur J Org Chem 2515
130. Trost BM, Schroeder GM (2000) J Org Chem 65:1569
131. Trost BM, Patterson DE, Hembre EJ (2001) Chem Eur J 7:3768
132. Trost BM, Toste FD (2000) J Am Chem Soc 122:11262
133. Loiseleur O, Elliott MC, von Matt P, Pfaltz A (2000) Helv Chim Acta 83:2287
134. Konno T, Hagata K, Ishihara T, Yamanaka H (2002) J Org Chem 67:1768
135. Weiß TD, Helmchen G, Kazmaier U (2002) Chem Commun 1270
136. Kazmaier U, Zumpe FL (2001) Eur J Org Chem 4067
137. Trost BM, Toste FD (1999) J Am Chem Soc 121:3543
138. Hilgraf R, Pfaltz A (1999) Synlett 11:1814
139. You S-L, Zhu X-Z, Luo Y-M, Hou X-L, Dai L-X (2001) J Am Chem Soc 123:7471
140. Evans PA, Brandt TA (1999) Org Lett 1:1563
141. Trost BM, Bunt RC, Lemoine RC, Calkins TL (2000) J Am Chem Soc 122:5968
142. Trost BM, Tang W, Schulte JL (2000) Org Lett 2:4013
143. Trost BM, McEachern EJ (1999) J Am Chem Soc 121:8649
144. Trost BM, Tsui H-C, Toste FD (2000) J Am Chem Soc 122:3534
145. Trost BM, Machacek MR (2002) Angew Chem Int Ed 41:4693
146. Trost BM, Fraisse PL, Ball ZT (2002) Angew Chem Int Ed 41:1059
147. Trost BM, Gunzner JL, Dirat O, Rhee YH (2002) J Am Chem Soc 124:10396
148. Trost BM, Asakawa N (1999) Synlett 1491
149. Glorius F, Pfaltz A (1999) Org Lett 1:141
150. Glorius F, Neuburger M, Pfaltz A (2001) Helv Chim Acta 84:3178
151. Malkov AV, Spoor P, Vinader V, Koovsk P (2001) Tetrahedron Lett 42:509

152. Belda O, Kaiser N-F, Bremberg U, Larhed M, Hallberg A, Moberg C (2000) J Org Chem 65:5868
153. Trost BM, Dogra K, Hachiya I, Emura T, Hughes DL, Krska S, Reamer RA, Palucki M, Yasuda N, Reider PJ (2002) Angew Chem Int Ed 41:1929
154. Palucki M, Um JM, Conlon DA, Yasuda N, Hughes DL, Mao B, Wang J, Reider PJ (2001) Adv Synth Catal 1:343
155. Kaiser N-FK, Bremberg U, Larhed M, Moberg C, Hallberg A (2000) Angew Chem Int Ed 39:3595
156. Krska SW, Hughes DL, Reamer RA, Mathre DJ, Sun Y, Trost BM (2002) J Am Chem Soc 124:12656
157. Hughes DL, Palucki M, Yasuda N, Reamer RA, Reider PJ (2002) J Org Chem 67:2762
158. Fuji K, Kinoshita N, Tanaka K, Kawabata T (1999) Chem Commun 2289
159. Bartels B, Helmchen G (1999) Chem Commun 741
160. Bartels B, García-Yebra C, Rominger F, Helmchen G (2002) Eur J Inorg Chem 2569
161. Ohmura T, Hartwig JF (2002) J Am Chem Soc 124:15164
162. Evans PA, Robinson JE (1999) Org Lett 1:1929
163. Evans PA, Robinson JE, Nelson JD (1999) J Am Chem Soc 121:6761
164. Evans PA, Robinson JE (2001) J Am Chem Soc 123:4609
165. Evans PA, Kennedy LJ (2001) J Am Chem Soc 123:1234
166. Song CE, Yang JW, Roh EJ, Lee S-G, Ahn JH, Han H (2002) Angew Chem Int Ed 41:3852
167. Trost BM, Pan Z, Zambrano J, Kujat C (2002) Angew Chem Int Ed 41:4691
168. Trost BM, Madsen R, Guile SD, Brown B (2000) J Am Chem Soc 122:5947
169. Okauchi T, Fujita K, Ohtaguro T, Ohshima S, Minami T (2000) Tetrahedron Asymmetry 11:1397
170. Trost BM, Dirat O, Dudash Jr J, Hembre EJ (2001) Angew Chem Int Ed 40:3658
171. Stragies R, Blechert S (2000) J Am Chem Soc 122:9584
172. Trost BM, Dudash J Jr, Hembre EJ (2001) Chem Eur J 7:1619
173. Trost BM, Lee CB (2001) J Am Chem Soc 123:3671
174. Trost BM, Lee CB (2001) J Am Chem Soc 123:3687
175. Trost BM, Schroeder GM (1999) J Am Chem Soc 121:6759
176. Trost BM, Schroeder GM, Kristensen J (2002) Angew Chem Int Ed 41:3492
177. You S-L, Hou X-L, Dai L-X, Zhu X-Z (2001) Org Lett 3:149
178. Brunel JM, Tenaglia A, Buono G (2000) Tetrahedron Asymmetry 11:3585
179. Trost DM, Ariza X (1999) J Am Chem Soc 121:10727
180. Trost BM, Dogra K (2002) J Am Chem Soc 124:7256
181. Nakoji M, Kanayama T, Okino T, Takemoto Y (2002) J Org Chem 67:7418

Supplement to Chapter 27
Allylation of C=O

Akira Yanagisawa

Department of Chemistry, Faculty of Science, Chiba University,
Inage, Chiba 263–8522, Japan
e-mail: ayanagi@scichem.s.chiba-u.ac.jp

Keywords: Allylation, Allylsilanes, Allylstannanes, Carbonyl compounds, Chiral Lewis acid catalysts, Chiral Lewis base catalysts

1   Introduction   97

2   Mechanism of Catalysis 100

3   Catalytic Allylation of Carbonyl Compounds 100
3.1   Allylation by Chiral Lewis Acid Catalysts 100
3.2   Allylation by Chiral Lewis Base Catalysts 103

4   Principal Alternative   105

References   105

1
Introduction

The chemistry of asymmetric allylation of carbonyl compounds has further progressed since the review in *Comprehensive Asymmetric Catalysis* [1] and plenty of papers including reviews [2, 3] on chiral catalysts for the reaction have since appeared. This chapter describes new examples of catalytic enantioselective allylation of carbonyl compounds with allylmetals in the presence of a catalytic amount of chiral Lewis acid or chiral Lewis base (Scheme 1). Compounds **1–36** [4–49] shown in Fig. 1 are the chiral catalysts reported since 1998, which have been used in the asymmetric allylation or propargylation of carbonyl compounds. Chiral compounds **37–40** [50–53], which have been utilized in the stoichiometric version, are also candidates for the chiral catalyst (Fig. 2).

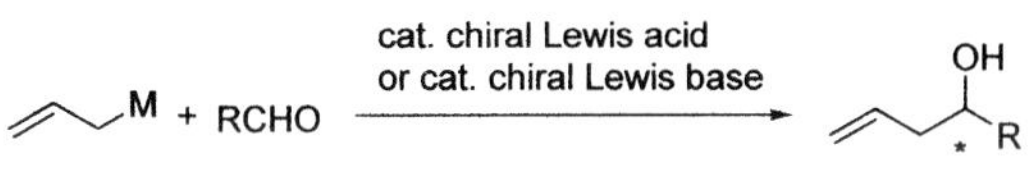

Scheme 1

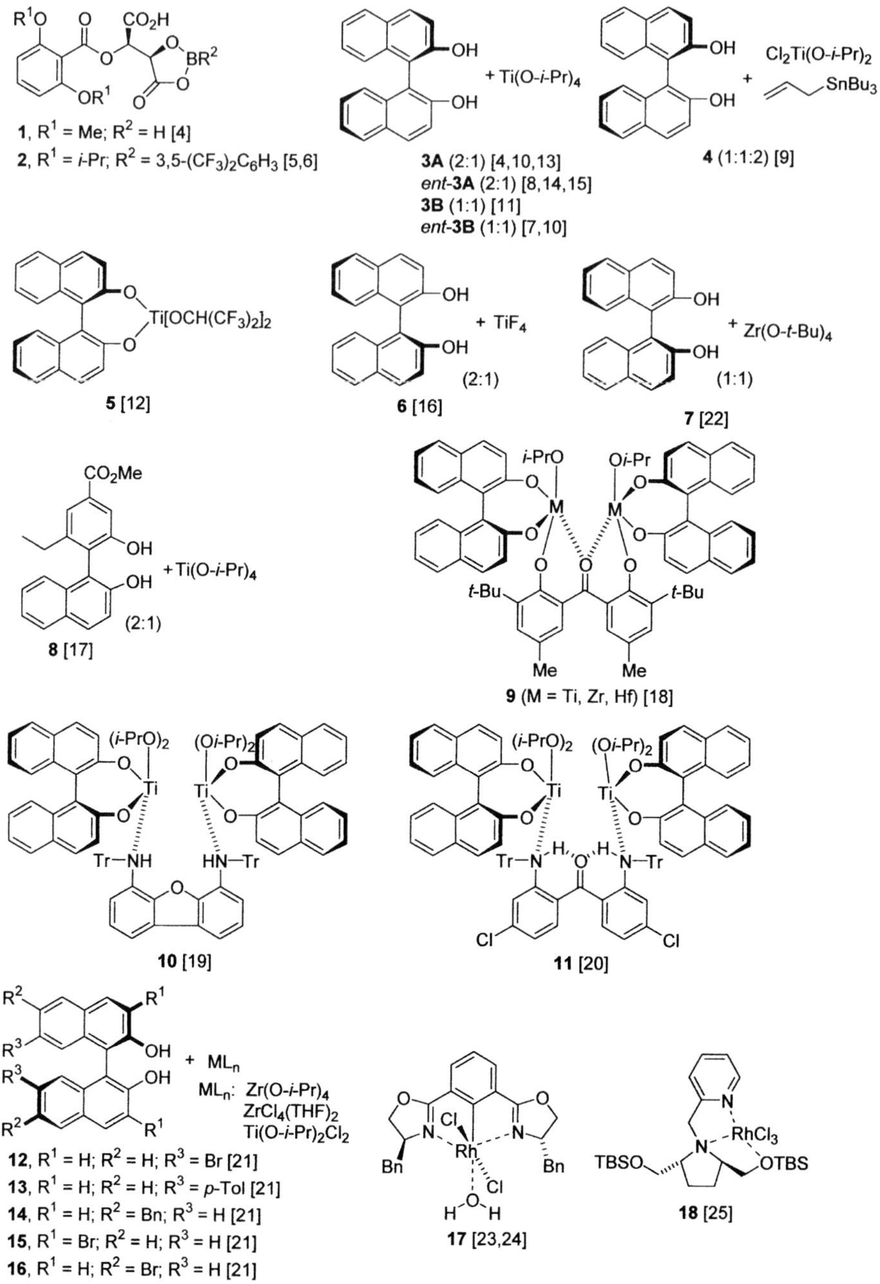

1, R^1 = Me; R^2 = H [4]
2, R^1 = *i*-Pr; R^2 = 3,5-(CF$_3$)$_2$C$_6$H$_3$ [5,6]

3A (2:1) [4,10,13]
ent-**3A** (2:1) [8,14,15]
3B (1:1) [11]
ent-**3B** (1:1) [7,10]

4 (1:1:2) [9]

5 [12]

6 [16]
(2:1)

7 [22]
(1:1)

8 [17]
(2:1)

9 (M = Ti, Zr, Hf) [18]

10 [19]

11 [20]

12, R^1 = H; R^2 = H; R^3 = Br [21]
13, R^1 = H; R^2 = H; R^3 = *p*-Tol [21]
14, R^1 = H; R^2 = Bn; R^3 = H [21]
15, R^1 = Br; R^2 = H; R^3 = H [21]
16, R^1 = H; R^2 = Br; R^3 = H [21]

ML$_n$: Zr(O-*i*-Pr)$_4$
ZrCl$_4$(THF)$_2$
Ti(O-*i*-Pr)$_2$Cl$_2$

17 [23,24]

18 [25]

Fig. 1

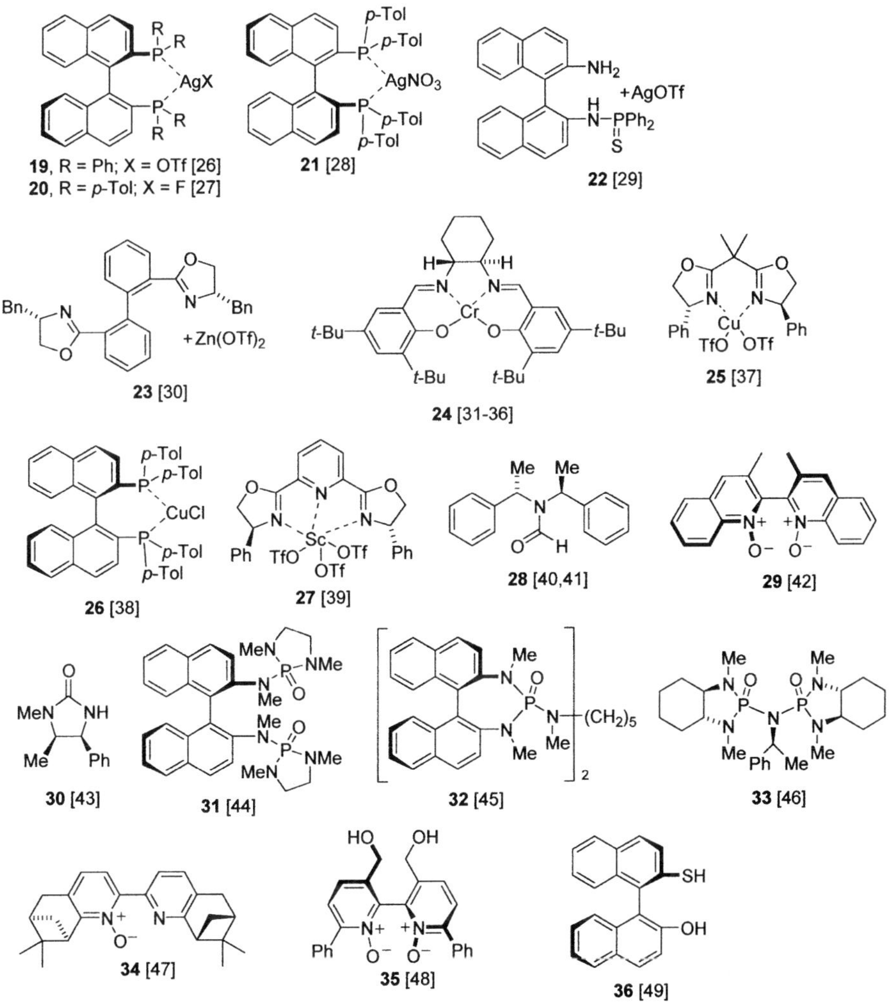
19, R = Ph; X = OTf [26]
20, R = p-Tol; X = F [27]
21 [28]
+AgOTf
22 [29]
+Zn(OTf)2
23 [30]
24 [31-36]
25 [37]
26 [38]
27 [39]
28 [40,41]
29 [42]
30 [43]
31 [44]
32 [45]
33 [46]
34 [47]
35 [48]
36 [49]

Fig. 1 (continued)

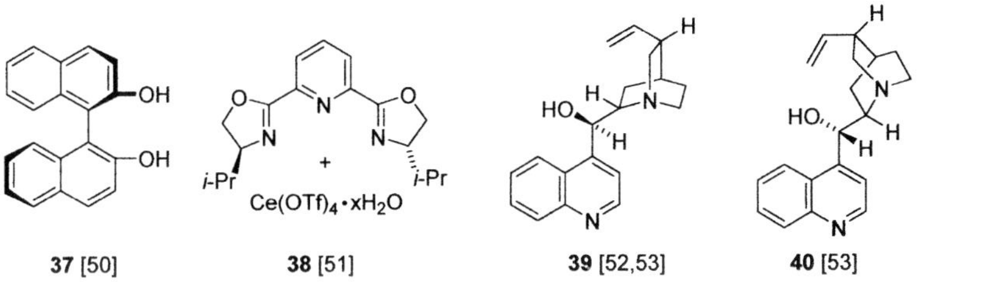
37 [50]
Ce(OTf)4·xH2O
38 [51]
39 [52,53]
40 [53]

Fig. 2

2
Mechanism of Catalysis

Several new methods for the asymmetric allylation of compounds have emerged; however, the catalytic mechanisms are essentially the same as those explained in Scheme 2 of the 1st edition.

3
Catalytic Allylation of Carbonyl Compounds

Catalytic asymmetric allylations of aldehydes or ketones are roughly classified into two methods, namely, those using chiral Lewis acid catalysts and those using chiral Lewis base catalysts. The former method uses less reactive allylsilanes or allylstannanes as the allyl source. The latter method requires allyltrichlorosilane or more reactive allylmetals. Both processes are applicable to the reactions with substituted allylmetal compounds or propargylation.

3.1
Allylation by Chiral Lewis Acid Catalysts

Chiral (acyloxy)borane (CAB) is known as an effective chiral Lewis acid catalyst for enantioselective allylation of aldehydes. Marshall applied the CAB complex **1** to the addition of crotylstannane to achiral aldehydes and found that the CAB catalyst gives higher *syn/anti* selectivity than BINOL/Ti catalysts in the reaction [4]. CAB complex **2** was utilized in asymmetric synthesis of chiral polymers using a combination of dialdehyde and bis(allylsilane) [5] or monomers possessing both formyl and allyltrimethylsilyl groups [6].

Chiral titanium complexes **3A**, *ent*-**3A**, **3B**, and *ent*-**3B**, first developed by Keck, were further extensively employed in catalytic asymmetric allylation reactions with various allylic stannanes and aldehydes [4, 7, 8, 10, 13]. If toluene or pentane is used as solvent, a highly enantioselective reaction can be accomplished even without molecular sieves [11]. Catalytic asymmetric allenylation of aldehydes with propargylic or allenylic stannanes has been also achieved by the catalyst *ent*-**3A** in the presence of synergetic reagent Et_2BS-*i*-Pr [14, 15]. Tagliavini has shown that chiral titanium reagent **4**, prepared in situ from a 1:1:2 mixture of BINOL, $TiCl_2(O$-*i*-Pr$)_2$, and allyltributyltin, can catalyze the asymmetric addition of tetraallyltin to ketones including aliphatic ketones which, however, afford lower enantiomeric excess [9]. (*R*)-BINOL-Ti complex **5** is also an efficient chiral catalyst for the addition of allylic tin reagents to aldehydes [12]. The TiF_4–BINOL **6**-catalyzed enantioselective allylsilylation of aldehydes has been further studied by Carreira et al. and they found that 2-naphthoate is the most effective protecting group for 2,2-dimethyl-3-hydroxypropionaldehyde in obtaining the corresponding nonracemic homoallylic alcohol, which is a key building block for polyketide synthesis [16]. Kurosu found that chiral BINOL–Zr complex **7** and BINOL–Ti complexes catalyze the asymmetric addition of allyltributyltin to aldehydes [22]. Chiral biaryl 8 is also a good chiral ligand for the asymmetric Ti-catalyzed allylation of aldehydes with allyltributyltin [17].

Maruoka has reported that chiral bimetallic Lewis acid catalysts 9–11, prepared from (S)-BINOL, M(O-i-Pr)$_4$ (M=Ti, Zr, Hf), and the corresponding spacer, strongly enhance the reactivity of aldehydes or ketones toward allyl transfer from allylstannanes [18–20]. For example, treatment of acetophenone (42) with tetraallyltin (41) in the presence of 30 mol% of the chiral bidentate Ti(IV) catalyst 10 provided the (S)-enriched homoallylic alcohol 43 in 95% yield with 90% ee (Scheme 2) [19]. A suggested reaction mechanism involves double activation of carbonyls owing to the simultaneous coordination of two Ti atoms to a carbonyl oxygen atom.

The effectiveness of various substituted BINOL ligands 12–16 in the Zr(IV)- or Ti(IV)-catalyzed enantioselective addition of allyltributyltin to aldehydes was also investigated by Spada and Umani-Ronchi [21]. The number of noteworthy examples of asymmetric allylation of carbonyl compounds utilizing optically active catalysts of late transition metal complexes has increased since 1999. Chiral bis(oxazolinyl)phenyl rhodium(III) complex 17, developed by Motoyama and Nishiyama, is an air-stable and water-tolerant asymmetric Lewis acid catalyst [23, 24]. Condensation of allylic stannanes with aldehydes under the influence of this catalyst results in formation of nonracemic allylated adducts with up to 80% ee (Scheme 3). In the case of the 2-butenyl addition reac-

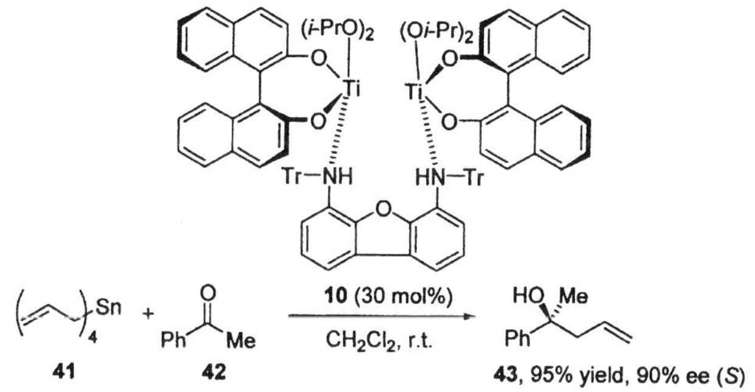

Scheme 2

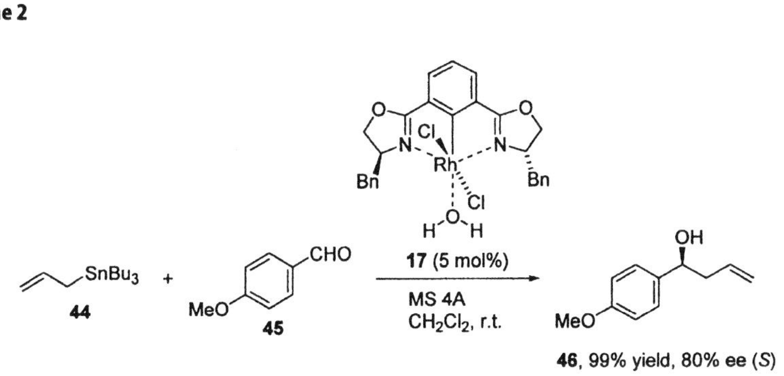

Scheme 3

tion, *anti* selectivity is observed regardless of the double bond geometry of the stannane [24].

The cationic chiral rhodium complex generated in situ from **18** and $AgBF_4$ is also a promising chiral catalyst for the reaction [25]. Our group reported full research details on BINAP–silver(I) triflate (**19**)-catalyzed asymmetric allylation of aldehydes with allylic stannanes [26]. The catalytic reaction has been further improved without reducing enantioselectivity by using allylic trimethoxysilanes as allylating agents and the *p*-Tol-BINAP·AgF catalyst (**20**), which is anticipated to activate the trimethoxysilane [27]. Remarkable γ and *anti* selectivities are observed for the reaction with crotylsilanes, irrespective of the *E*/*Z* configuration at the double bond. For instance, addition of *E*-enriched crotyltrimethoxysilane (**47**, *E*/*Z*=83/17) to benzaldehyde (**48**) in the presence of a catalytic amount of (*R*)-BINAP·AgF in MeOH at -20°C to room temperature provides almost quantitatively the γ-adducts **49** and **50** with an *anti*/*syn* ratio of 92:8. The *anti* isomer **49** has 96% ee (Scheme 4).

A similar chiral silver(I) catalyst **21** was applied to the asymmetric addition of allyltributyltin to various aldehydes in an aqueous medium [28]. Shi et al. have shown that chiral silver complex **22**, prepared from chiral bidentate phosphoramide and AgOTf, is also an effective chiral catalyst for the allylation [29]. Chiral bis(oxazoline) ligands have found widespread use in asymmetric reactions catalyzed by chiral metal complexes, and C_2-symmetric chiral bis(oxazoline)–$Zn(OTf)_2$ complex **23** has been applied to catalytic enantioselective allylation of aldehydes with allyltributyltin (**44**); however, satisfactory enantioselectivity was not observed [30].

Cozzi and Umani-Ronchi have achieved the first catalytic enantioselective allylation reaction (Nozaki–Hiyama reaction) of aldehydes using a catalytic amount of chiral [Cr(Salen)] complex **24** as chiral Lewis acid catalyst [31, 32]. They found that when allyl chloride (**51**) was reacted with cyclohexanecarbaldehyde (**52**) in the presence of 10 mol% of the chiral Cr catalyst **24**, Mn, and Me_3SiCl followed by acidic work-up, the adduct **53** was obtained with 89% ee (Scheme 5). The decisive points of this catalytic redox cycle are the employment of Mn as the stoichiometric reductant and Me_3SiCl as the scavenger. In general, by-products derived from pinacol coupling or reduction of aldehydes are obtained in the reaction. In the addition of crotyl bromide to benzaldehyde, by using a 10 mol% excess of the chiral Salen ligand, complete inversion of the diastereoselectivity (*anti* to *syn*) was observed [33, 34]. The catalytic diastereose-

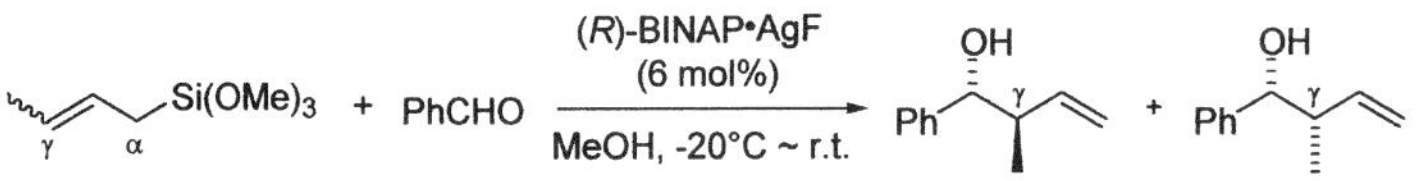

47 (*E*/*Z* = 83/17) **48** **49**, *anti* (1*R*,2*R*) **50**, *syn* (1*R*,2*S*)

77% combined yield
anti/*syn* = 92 (96% ee)/8

Scheme 4

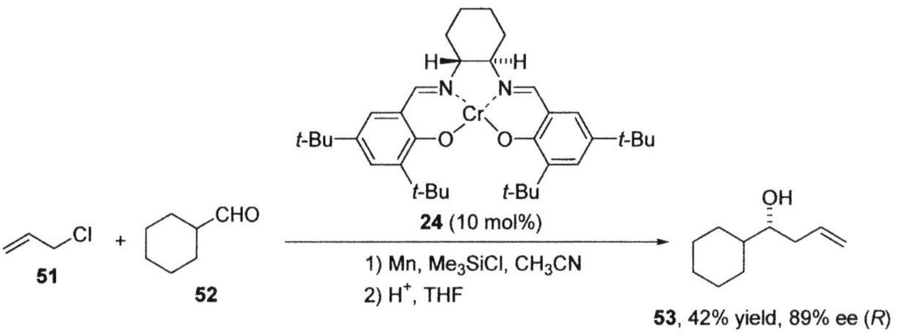

Scheme 5

lective and enantioselective reaction of 1,3-dichloropropanes to aromatic aldehydes has further been applied to the synthesis of optically active1,2-*syn*-chlo-
rohydrins, key precursors of *cis*-vinyl epoxides [35]. Asymmetric propargyla-
tions of aldehydes have been also accomplished by the chromium catalyst **24**
[36].

Jørgensen et al. reported that C_2-symmetric bis(oxazoline)–copper(II) complex **25** also acts as chiral Lewis acid catalyst for a reaction of allylic stannane
with ethyl glyoxylate [37]. Meanwhile, *p*-Tol-BINAP–CuCl complex **26** was
shown to be a promising chiral catalyst for a catalytic enantioselective allylation of ketones with allyltrimethoxysilane under the influence of the TBAT catalyst [38]. Evans and coworkers have developed (*S,S*)-Ph-pybox·Sc(OTf)$_3$ complex **27** as a new chiral Lewis acid catalyst and shown that this scandium catalyst
promotes enantioselective addition reactions of allenyltrimethylsilanes to ethyl
glyoxylate [39]. But, when the silicon substituents become bulkier, nonracemic
dihydrofurans are predominantly obtained as products of [3+2] cycloaddition.

3.2
Allylation by Chiral Lewis Base Catalysts

Many noticeable examples of chiral Lewis base catalyzed allylation of carbonyl compounds have also appeared. Iseki and coworkers published a full paper on enantioselective addition of allyl- and crotyltrichlorosilanes to aliphatic aldehydes catalyzed by a chiral formamide **28** in the presence of HMPA as
an additive [41]. This method was further applied to asymmetric allenylation
of aliphatic aldehydes with propargyltrichlorosilane [40]. Nakajima and Hashimoto have demonstrated the effectiveness of (*S*)-3,3′-dimethyl-2,2′-biquinoline
N,N′-dioxide (**29**) as a chiral Lewis base catalyst for the allylation of aldehydes
[42]. In the reaction of (*E*)-enriched crotyltrichlorosilane (**54E**, *E:Z*=97:3) with
benzaldehyde (**48**), γ-allylated *anti*-homoallylic alcohol **55** was obtained exclusively with high ee while the corresponding *syn*-adduct was formed from its *Z*
isomer **54Z** (*E:Z*=1:99) (Scheme 6). Catalytic amounts of chiral urea **30** also promote the asymmetric reaction in the presence of a silver(I) salt, although the
enantioselectivity is low [43].

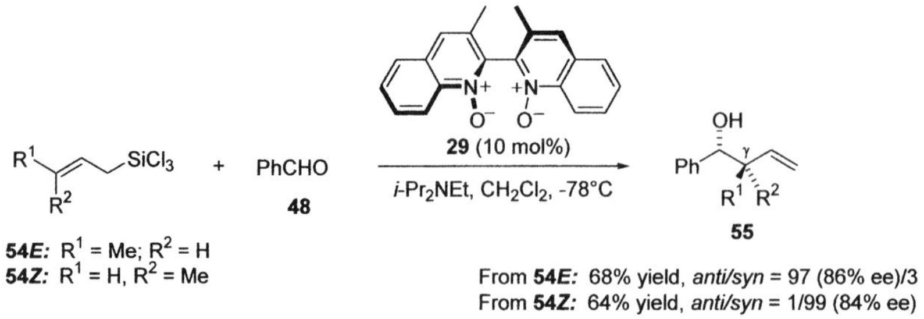

Scheme 6

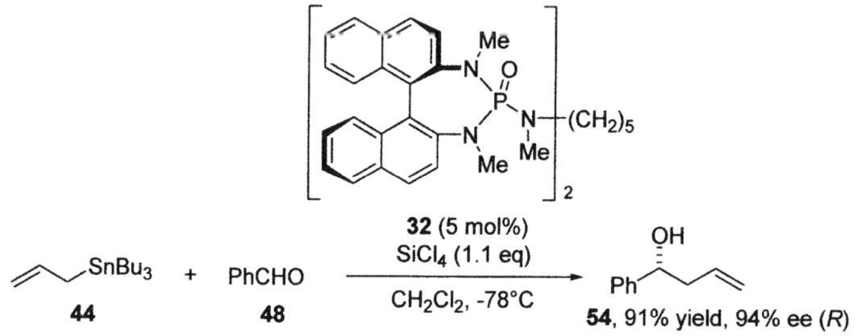

From **54E:** 68% yield, *anti/syn* = 97 (86% ee)/3
From **54Z:** 64% yield, *anti/syn* = 1/99 (84% ee)

54E: R^1 = Me; R^2 = H
54Z: R^1 = H, R^2 = Me

Scheme 7

Denmark et al. have developed binaphthyldiamine-derived bisphosphoramide **31** as a new chiral Lewis base catalyst for enantioselective allylation of aldehydes using allyltrichlorosilane, with the knowledge that a four- or five-unit tether is necessary for the bisphosphoramide to achieve highly asymmetric induction [44]. In fact, they observed higher enantioselectivity in the reaction of benzaldehyde promoted by the bisphosphoramide **31** than that shown by the previously reported chiral phosphoramides (compound **14** in the original review [1]). They have further improved this asymmetric allyl transfer process and have found that silicon tetrachloride is activated by a chiral Lewis base to catalyze the allylation of aldehydes with allyltributyltin (**44**) [45]. For example, the binaphthyl bisphosphoramide possessing a five-methylene linker **32** gave the allylated product **54** in 91% yield and 94% ee with only 5 mol % of this chiral Lewis base catalyst and 1.1 equivalents of SiCl$_4$ (Scheme 7). This chiral phosphoramide–SiCl$_4$ system was also applied to the addition of allenyltributylstannane to aldehydes [45].

Chiral bidentate imidodiphosphoric tetramide **33** [46], chiral 2,2′-bipyridine-type *N*-monoxide (PINDOX) **34** [47], and chiral 3,3′-bis(hydroxymethyl)-6,6′-diphenyl-2,2′-bipyridine *N,N*′-dioxide (**35**) [48] have been also developed as efficient chiral Lewis base catalysts for the asymmetric allylation of aldehydes with allyltrichlorosilane.

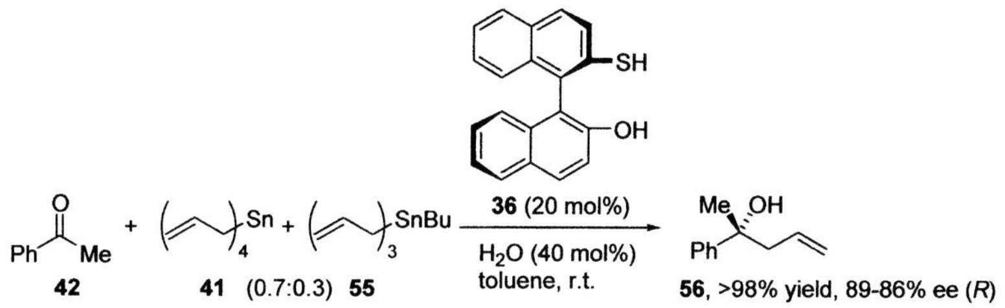

Scheme 8

4
Principal Alternative

Chiral Brønsted acids can also promote the asymmetric addition of allylic tin reagents to carbonyl compounds. Baba and coworkers have found that a stoichiometric amount of (R)-BINOL (**37**) acts a chiral promoter for the allylation of unactivated ketones with tetraallyltin and in the presence of MeOH, the corresponding nonracemic tertiary homoallylic alcohols are obtained with up to 60% ee [50]. Later, Woodward et al. improved this process and achieved a catalytic enantioselective allylation of aryl ketones by employing (R)-monothiobinaphthol **36** as a chiral Brønsted catalyst [49]. For instance, in the presence of 20 mol% of the chiral acid **36** and 40 mol% of H_2O in toluene, acetophenone (**42**) was allylated by a 0.7:0.3 mixture of tetraallyltin (**41**) and butyltriallyltin (**55**) to give the (R)-enriched allylated product **56** almost quantitatively with 89–86% ee (Scheme 8).

Several methods promoted by a stoichiometric amount of chiral Lewis acid **38** [51] or chiral Lewis bases **39** [52, 53] and **40** [53] have been developed for enantioselective indium-mediated allylation of aldehydes and ketones by the Loh group. A combination of a chiral trimethylsilyl ether derived from norpseudoephedrine and allyltrimethylsilane is also convenient for synthesis of enantiopure homoallylic alcohols from ketones [54, 55]. Asymmetric carbonyl addition by chirally modified allylic metal reagents, to which chiral auxiliaries are covalently bonded, is also an efficient method to obtain enantiomerically enriched homoallylic alcohols and various excellent chiral allylating agents have been developed: for example, (1S,2S)-pseudoephedrine- and (1R,2R)-cyclohexane-1,2-diamine-derived allylsilanes [56], polymer-supported chiral allylboron reagents [57], and a bisoxazoline-modified chiral allylzinc reagent [58]. An allyl transfer reaction from a chiral crotyl donor opened a way to highly enantioselective and α-selective crotylation of aldehydes [59–62]. Enzymatic routes to enantioselective allylation of carbonyl compounds have still not appeared.

References

1. Yanagisawa A, Yamamoto H (1999) In: Jacobsen EN, Pfaltz A, Yamamoto H (eds) Comprehensive asymmetric catalysis III, chap 34.2. Springer, Berlin Heidelberg New York, p 1295

2. Denmark SE, Almstead NG (2000) In: Otera J (ed) Modern Carbonyl chemistry, Chap 10. Wiley-VCH, Weinheim, p 299
3. Chemler SR, Roush WR (2000) In: Otera J (ed) Modern carbonyl chemistry, Chap 11. Wiley-VCH, Weinheim, p 403
4. Marshall JA, Palovich MR (1998) J Org Chem 63:4381
5. Kumagai T, Itsuno S (2001) Tetrahedron Asymmetry 12:2509
6. Itsuno S, Kumagai T (2002) Synthesis 941
7. Keck GE, Krishnamurthy D (1998) Org Synth 75:12
8. Keck GE, Yu T (1999) Org Lett 1:289
9. Casolari S, D'Addario D, Tagliavini E (1999) Org Lett 1:1061
10. Zimmer R, Hain U, Berndt M, Gewald, R, Reissig HU (2000) Tetrahedron Asymmetry 11:879
11. Doucet H, Santelli M (2000) Tetrahedron Asymmetry 11:4163
12. Yu C-M, Lee J-Y, So B, Hong J (2002) Angew Chem Int Ed 41:161
13. Keck GE, Covel JA, Schiff T, Yu T (2002) Org Lett 4:1189
14. Yu C-M, Yoon S-K, Baek K, Lee J-Y (2002) Angew Chem Int Ed 37:2392
15. Yu C-M, Yoon S-K, Lee S-J, Lee J-Y, Kim SS (1998) Chem Commun 2749
16. Bode JW, Gauthier DR Jr, Carreira EM (2001) Chem Commun 2560
17. Brenna E, Scaramelli L, Serra S (2000) Synlett 357
18. Hanawa H, Kii S, Asao N, Maruoka K (2000) Tetrahedron Lett 41:5543
19. Hanawa H, Kii S, Maruoka K (2001) Adv Synth Catal 343:57
20. Kii S, Maruoka K (2001) Tetrahedron Lett 42:1935
21. Bandin M, Casolari S, Cozzi PG, Proni G, Schmohel E, Spada GP, Tagliavini E, Umani-Ronchi A (2000) Eur J Org Chem 491
22. Kurosu M, Lorca M (2002) Tetrahedron Lett 43:1765
23. Motoyama Y, Narusawa H, Nishiyama H (1999) Chem Commun 131
24. Motoyama Y, Okano M, Narusawa H, Makihara N, Aoki K, Nishiyama H (2001) Organometallics 20:1580
25. Shi M, Lei G-X, Masaki Y (1999) Tetrahedron Asymmetry 10:2071
26. Yanagisawa A, Nakashima H, Nakatsuka Y, Ishiba A, Yamamoto H (2001) Bull Chem Soc Jpn 74:1129
27. Yanagisawa A, Kageyama H, Nakatsuka Y, Asakawa K, Matsumoto Y, Yamamoto H (1999) Angew Chem Int Ed 38:3701
28. Loh T-P, Zhou J-R (2000) Tetrahedron Lett 41:5261
29. Shi M, Sui W-S, Masaki Y (2000) Tetrahedron Asymmetry 11:773
30. Imai Y, Zhang W, Kida T, Nakatsuji Y, Ikeda I (2000) J Org Chem 65:3326
31. Bandini M, Cozzi PG, Umani-Ronchi A (2002) Chem Commun 919
32. Bandini M, Cozzi PG, Melchiorre P, Umani-Ronchi A (1999) Angew Chem Int Ed 38:3357
33. Bandini M, Cozzi PG, Umani-Ronchi A (2000) Angew Chem Int Ed 39:2327
34. Bandini M, Cozzi PG, Umani-Ronchi A (2001) Tetrahedron 57:835
35. Bandini M, Cozzi PG, Melchiorre P, Morganti S, Umani-Ronchi A (2001) Org Lett 3:1153
36. Bandini M, Cozzi PG, Melchiorre P, Tino R, Umani-Ronchi A (2001) Tetrahedron Asymmetry 12:1063
37. Gathergood N, Jørgensen KA (1999) Chem Commun 1869
38. Yamasaki S, Fujii K, Wada R, Kanai M, Shibasaki M (2002) J Am Chem Soc 124:6536
39. Evans DA, Sweeney ZK, Rovis T, Tedrow JS (2001) J Am Chem Soc 123:12095
40. Iseki K, Kuroki Y, Kobayashi Y (1998) Tetrahedron Asymmetry 9:2889
41. Iseki K, Mizuno S, Kuroki Y, Kobayashi Y (1998) Tetrahedron 55:977
42. Nakajima M, Saito M, Shiro M, Hashimoto S (1998) J Am Chem Soc 120:6419
43. Chataigner I, Piarulli U, Gennari C (1999) Tetrahedron Lett 40:3633
44. Denmark SE, Fu J (2000) J Am Chem Soc 122:12021
45. Denmark SE, Wynn T (2001) J Am Chem Soc 123:6199
46. Hellwig J, Belser T, Müller JFK (2001) Tetrahedron Lett 42:5417

47. Malkov AV, Orsini M, Pernazza D, Muir KW, Langer V, Meghani P, Kocovsky P (2002) Org Lett 4:1047
48. Shimada T, Kina A, Ikeda S, Hayashi T (2002) Org Lett 4:2799
49. Cunningham A, Woodward S (2002) Synlett 43
50. Yasuda M, Kitahara N, Fujibayashi T, Baba A (1998) Chem Lett 743
51. Loh T-P, Zhou J-R (1999) Tetrahedron Lett 40:9115
52. Loh T-P, Zhou J-R, Li X-R (1999) Tetrahedron Lett 40:9333
53. Loh T-P, Zhou J-R, Yin Z (1999) Org Lett 1:1855
54. Tietze LF, Schiemann K, Wegner C, Wulff C (1998) Chem Eur J 4:1862
55. Tietze LF, Völkel L, Wulff C, Weigand B, Bittner C, McGrath P, Johnson K, Schäfer M (2001) Chem Eur J 7:1304
56. Kinnaird JWA, Ng PY, Kubota K, Wang X (2002) J Am Chem Soc 124:7920
57. Itsuno S, Watanabe K, El-Shehawy AA (2001) Adv Synth Catal 343:89
58. Nakamura M, Hirai A, Sogi M, Nakamura E (1998) J Am Chem Soc 120:5846
59. Nokami J, Ohga M, Nakamoto H, Matsubara T, Hussain I, Kataoka K (2001) J Am Chem Soc 123:9168
60. Hussain I, Komasaka T, Ohga M, Nokami J (2002) Synlett 640
61. Loh T-P, Hu Q-Y, Chok Y-K, Tan K-T (2001) Tetrahedron Lett 42:9277
62. Loh T-P, Lee C-LK, Tan K-T (2002) Org Lett 4:2985

Supplement to Chapter 31.1
Conjugate Addition of Organometallics to Activated Olefins

Kiyoshi Tomioka

Graduate School of Pharmaceutical Sciences, Kyoto University,
Yoshida, Sakyo-ku, 606–8501, Kyoto, Japan
e-mail: tomioka@pharm.kyoto-u.ac.jp

Keywords: Organozinc, Organolithium, Grignard reagent, Copper, Conjugate addition, Nucleophilic addition, Alkylation, Activated olefin, Enone, Nitroalkene, β-Substituted carbonyl compound, Ligand, Phosphorus, Phosphine

1	**Introduction**	109
2	**Reaction of Organozinc Reagents**	110
2.1	Copper-Catalyzed Reaction with Cyclic Enones by Using External Chiral Ligands	110
2.2	Nickel- and Copper-Catalyzed Reaction with Acyclic Enones by Using External Chiral Ligands	114
2.3	Copper-Catalyzed Reaction with Nitroalkenes by Using External Chiral Ligands	116
3	**Rhodium-Catalyzed Arylation Reaction of Arylmetallic Reagents by Using External Chiral Ligands**	117
4	**Radical Reaction by Using External Chiral Ligands**	119
5	**Reaction of Organolithium Reagents by Using External Chiral Ligands**	120
6	**Reaction of Chiral Organometallic Reagents**	122
7	**Principal Alternatives**	122
	References	122

1
Introduction

Considerable progress has been made in the asymmetric conjugate addition of organometallic reagents to electrophilically activated olefins over the last six years (1998–2003) [1]. In this chapter the recent progress in the enantioselective conjugate addition reaction of organometallic reagents to electrophilically activated olefins under the control of chiral catalysts or external chiral ligands [2] will be summarized. The Michael reaction of the active methylene compounds is not included in this chapter.

2
Reaction of Organozinc Reagents

The copper-catalyzed conjugate addition reaction of organozinc reagents has reached a prominent level of success by the development of sophisticated chiral phosphorus ligands. Nickel catalysts have also been surveyed by several groups.

2.1
Copper-Catalyzed Reaction with Cyclic Enones by Using External Chiral Ligands

Asymmetric conjugate addition of organozinc reagents to enones in the presence of a chiral ligand is a rapidly developing and exciting area in the field of conjugate addition chemistry. The chiral phosphoramidite 1, developed by Feringa, is the best prototype ligand for copper triflate-catalyzed conjugate addition of organozinc reagents to cyclohexenone, which is based on the use of binaphthol and a chiral amine, to give 3-ethylcyclohexanone in satisfactorily high enantioselectivity [3]. A wide range of modifications of the prototype ligand 1 have been provided by the many groups summarized in Table 1 [3–25].

The phosphoramidites 2–6 are typical modified ligands that give an ethyl adduct in over 90% ee [4–8]. The advantage of organozinc reagents is illustrated by the use of a functionalized zinc reagent in the synthesis of 24 in 87% ee by using ligand 3 [5]. Polymer-bound phosphoramidite 6 has useful applications [8]. Dimethylzinc was also shown to be applicable as a source of a methyl group by using 5 in the preparation of muscone 25 in 95% ee [7]. More extended use of a binaphthol unit as a constituent of the phosphoramidite has appeared as illustrated by ligands 7–10, which give ethylcyclohexanone in 77–96% ee [9–12].

It is remarkable that a biphenyl unit in combination with a chiral amine 11 plays a role of an alternate binaphthol, inducing atropisomerism and giving ethylcyclohexanone in 96% ee [13]. Alexakis also succeeded in his search for efficient copper salts and solvents and reported the combination of CuTC (copper thiophene-2-carboxylate) and ether as a suitable choice. It is also noteworthy that trapping of the resulting zinc enolate with TMSOTf gave a TMS enolate 26, which is a useful intermediate for further synthetic manipulation [26]. Tandem

Table 1 Catalytic asymmetric addition of diethylzinc to cyclohexenone in the presence of 1–23

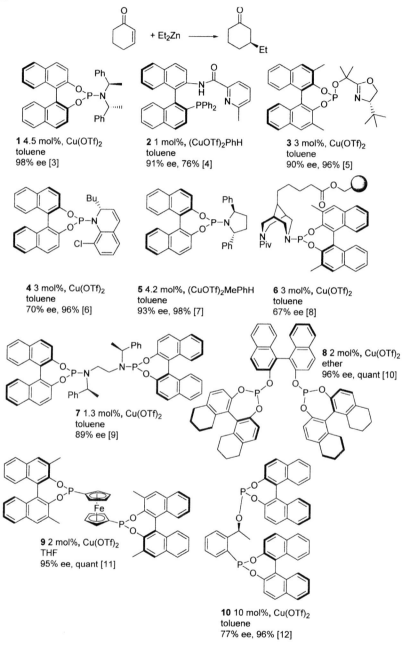

1 4.5 mol%, Cu(OTf)₂
toluene
98% ee [3]

2 1 mol%, (CuOTf)₂PhH
toluene
91% ee, 76% [4]

3 3 mol%, Cu(OTf)₂
toluene
90% ee, 96% [5]

4 3 mol%, Cu(OTf)₂
toluene
70% ee, 96% [6]

5 4.2 mol%, (CuOTf)₂MePhH
toluene
93% ee, 98% [7]

6 3 mol%, Cu(OTf)₂
toluene
67% ee [8]

7 1.3 mol%, Cu(OTf)₂
toluene
89% ee [9]

8 2 mol%, Cu(OTf)₂
ether
96% ee, quant [10]

9 2 mol%, Cu(OTf)₂
THF
95% ee, quant [11]

10 10 mol%, Cu(OTf)₂
toluene
77% ee, 96% [12]

Table 1 (continued)

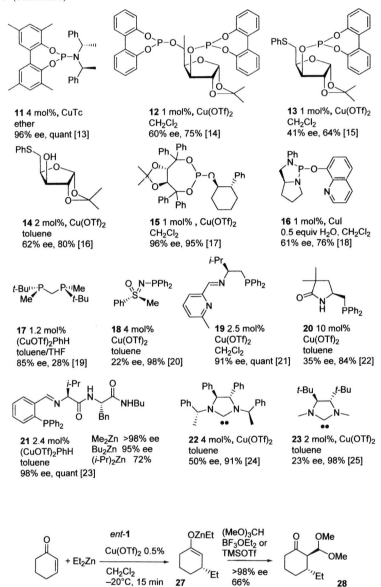

11 4 mol%, CuTc
ether
96% ee, quant [13]

12 1 mol%, Cu(OTf)$_2$
CH$_2$Cl$_2$
60% ee, 75% [14]

13 1 mol%, Cu(OTf)$_2$
CH$_2$Cl$_2$
41% ee, 64% [15]

14 2 mol%, Cu(OTf)$_2$
toluene
62% ee, 80% [16]

15 1 mol% , Cu(OTf)$_2$
CH$_2$Cl$_2$
96% ee, 95% [17]

16 1 mol%, CuI
0.5 equiv H$_2$O, CH$_2$Cl$_2$
61% ee, 76% [18]

17 1.2 mol%
(CuOTf)$_2$PhH
toluene/THF
85% ee, 28% [19]

18 4 mol%
Cu(OTf)$_2$
toluene
22% ee, 98% [20]

19 2.5 mol%
Cu(OTf)$_2$
CH$_2$Cl$_2$
91% ee, quant [21]

20 10 mol%
Cu(OTf)$_2$
toluene
35% ee, 84% [22]

21 2.4 mol%
(CuOTf)$_2$PhH
toluene
98% ee, quant [23]

Me$_2$Zn >98% ee
Bu$_2$Zn 95% ee
(i-Pr)$_2$Zn 72%

22 4 mol%, Cu(OTf)$_2$
toluene
50% ee, 91% [24]

23 2 mol%, Cu(OTf)$_2$
toluene
23% ee, 98% [25]

Scheme 1

conjugate addition to cyclohexenone in methylene chloride and electrophilic trapping of the zinc enolate **27** with an acetal to **28** was one notable achievement (Scheme 1) [27].

Discrimination of enantiotopic olefins **29** was demonstrated by Alexakis; thus, cyclohexadienone **29** gave the diethylated adduct **31** with high enantiose-lectivity and diastereoselectivity via the intermediate **30** (Scheme 2) [28].

The introduction of sugar cores in **12** and **13** provided new ligands depart-ing from a simple combination of binaphthol and chiral amine with moderate success [14, 15]. It is interesting that the sulfur-containing ligand **14** gave ethyl-cyclohexanone in 62% ee [16]. Taddol-derived phosphoramidite **15** was found to effect the addition reaction to give 96% ee [17]. Bicyclo[3.3.0]-type ligand **16** was moderately effective with use of CuI and gave 61% ee [18].

Bisphosphine **17** was the first bisphosphine ligand to be used for the copper-catalyzed organozinc addition reaction and gave 85% ee [19]. The simple phos-phine **18** is a rare example of a chiral sulfone-derived ligand [20]. Valine-derived ligand **19** is effective and gave 91% ee [21]. The amidophosphine **20** gave only 35% selectivity [22].

Hoveyda's phosphine **21** is an exciting recent development and gave quite high levels of generality with respect to organozinc reagents and quite high enanti-oselectivities [23]. Hoveyda succeeded in the conjugate addition of dimethylz-inc to **32** and subsequent alkylation of zinc enolate **33** to provide **34** in 97% ee and 80% yield (Scheme 3).

The two chiral carbene ligands **22** and **23** are of interest, although the selec-tivities achieved were not high [24, 25].

The copper-catalyzed addition of butyl Grignard reagent to cyclohexenone was reported by using selenium-oxazoline-based ligand **35** which gave butylcy-clohexanone **36** in 62% ee (Scheme 4) [24].

The conjugate addition of diethylzinc to 5,6-dihydropyran-2-one **37** to give lactone **38** in 92% ee is a rare successful example of the use of the bridged bi-sphosphoramidite **39** developed by Chan (Scheme 5) [30].

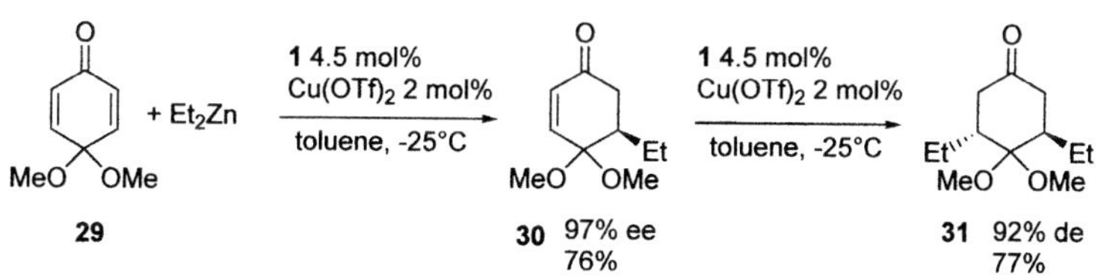

Scheme 2

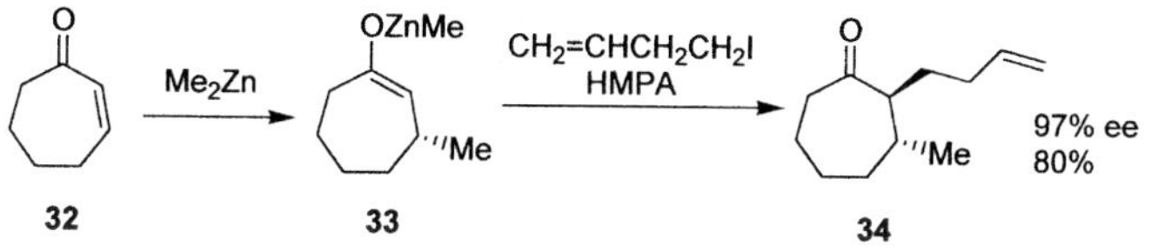

Scheme 3

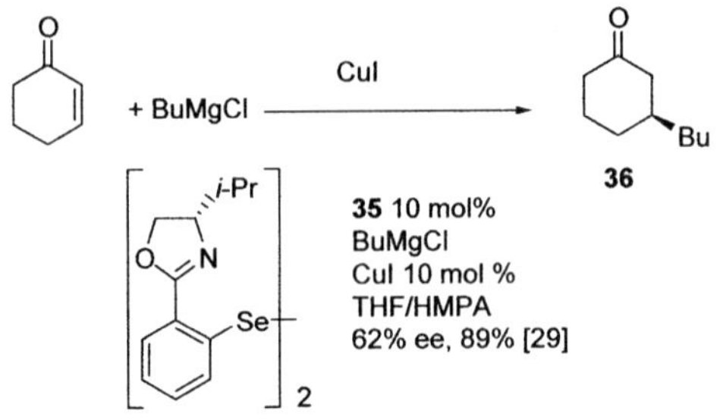

Scheme 4

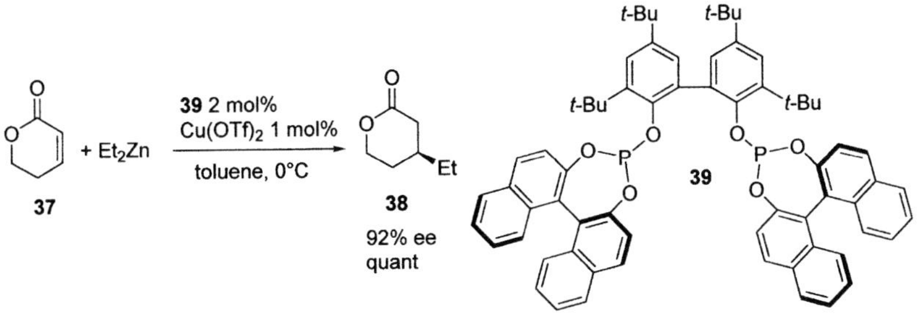

Scheme 5

2.2
Nickel- and Copper-Catalyzed Reaction with Acyclic Enones by Using External Chiral Ligands

Asymmetric conjugate addition of organozinc reagents to acyclic enones has been one of the topics of interest in recent conjugate addition chemistry as summarized in Table 2. The phosphoramidite **2** was effective toward chalcone **40** to give an ethyl adduct **41** in 96% ee [4]. The chiral oxazoline-based phosphaferrocene **42** was also a good ligand for copper-catalyzed reaction to give **41** in 87% ee [31].

Chiral bidentate alcohol–nickel catalysts **43–46** in place of copper catalyst have been a long-standing target and give **41** in up to 90% ee in acetonitrile or propionitrile [35]. The binaphthol-based phosphoramidite **47** was apparently a good ligand for copper and gave **41** in good ee [36].

A binaphthol-based sulfur-containing ligand **48** was effective in THF for a methyl ketone-type enone **49** and gave adduct **50** in 72% ee (Scheme 6) [37]. The phosphine **53**, developed by Hoveyda, was an epoch-making ligand for use with methyl ketone-type enones. For example, **51** gave the ethyl adduct **52** in 94% ee by using **53** and copper(I) [38]. The **56**-catalyzed cross metathesis reaction of olefin **54** with methyl vinyl ketone **55** gave the enone substrate **57** that underwent conjugate addition of diethylzinc to directly afford the cyclization product **58** in 85% ee and 78% yield [38].

Table 2 Catalytic asymmetric addition of diethylzinc to chalcone in the presence of 2 and 42–47

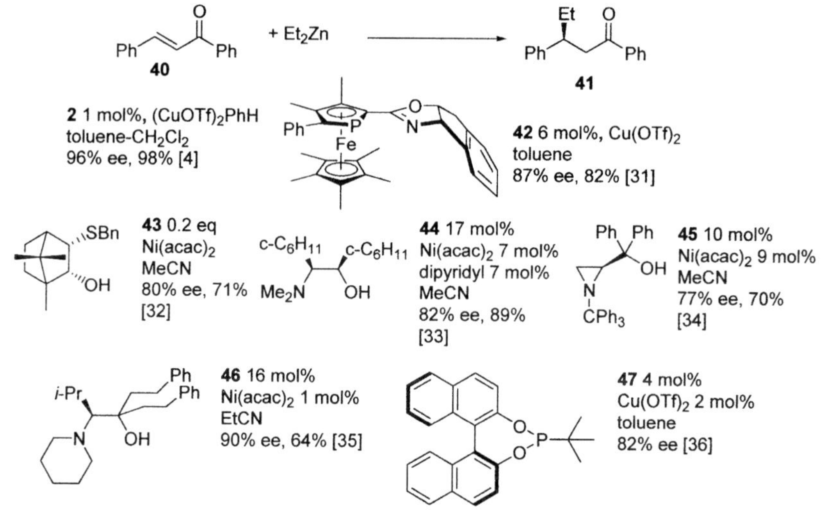

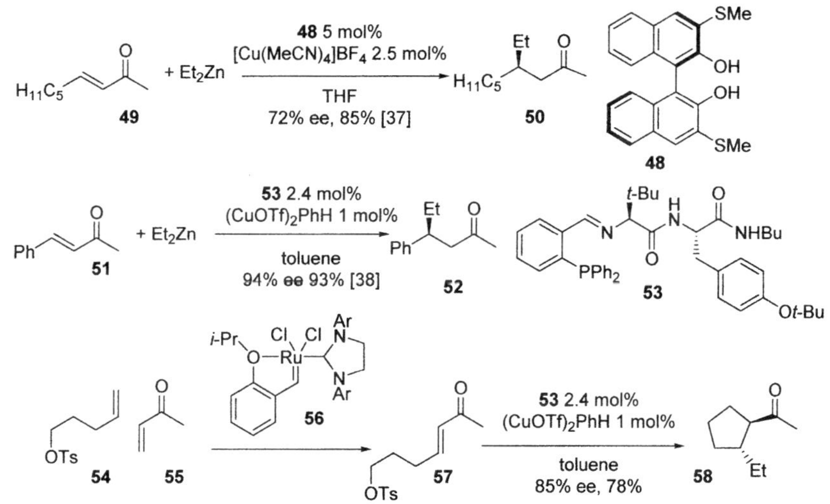

Scheme 6

Alkylidenemalonate **59** was a substrate in the Taddol-based phosphoramidite **61**-mediated addition of diethyl zinc to afford the ethyl adduct **60** in 73% ee [39]. The unsaturated imine **62** was a substrate in an amidephosphine **64** copper-catalyzed ethylation to afford **63** in moderate ee after hydrolysis (Scheme 7) [40].

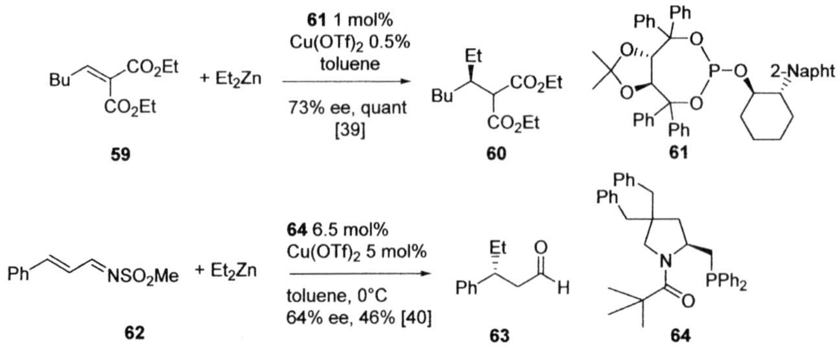

Scheme 7

Table 3 Asymmetric conjugate addition to **65**

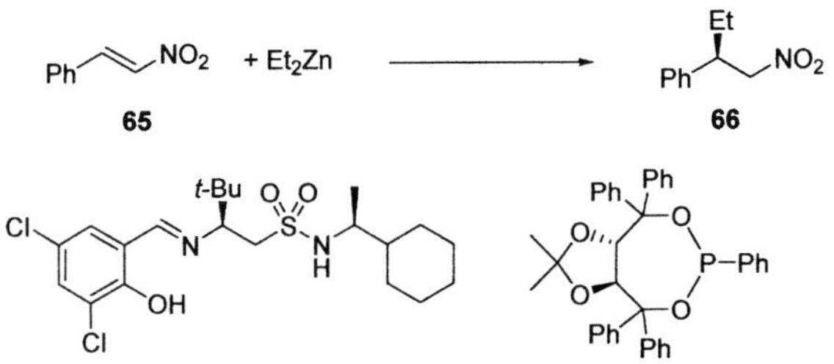

67 5.5 mol%, $Cu(OTf)_2$ 5 mol%
toluene, -20°C
58% ee, 55% [41]

68 1%, $Cu(OTf)_2$ 0.5%
toluene -30°C
81% ee, quant [42]

2.3 Copper-Catalyzed Reaction with Nitroalkenes by Using External Chiral Ligands

The asymmetric conjugate addition of organozinc reagents to nitroalkenes has also been an area of impressive progress. The phenolic ligand **67** shows moderate enantioselectivity and gave **66** in 58% ee and 55% yield (Table 3) [41,42]. Improvement was achieved by using the Taddol-based phosphoramidite **68** to afford **66** in 81% ee and quantitative yield, although the reaction of the cyclic nitroalkene **69** gave **70** in only 60% ee (Table 4) [13,42–44]. The Feringa phosphoramidite **1** again shows a high level of efficiency and gave **70** in 90% ee [43]. The Alexakis ligand **11** behaved in a satisfactory manner to give **70** in 95% ee with a combination of copper naphthoate as a copper source in an ether solvent [13]. The best ligand was again provided by Hoveyda and 96% ee and 92% yield were achieved by the use of **71** [44].

The binaphthol-based Feringa phosphoramidite **1** and its more sterically demanding modification **76** were goods ligands for the reaction with nitroalkenes **72**, **74**, and **77** and gave enantioselectivities of over 92% (Scheme 8) [45,46]. The

Table 4 Asymmetric conjugate addition to **69**

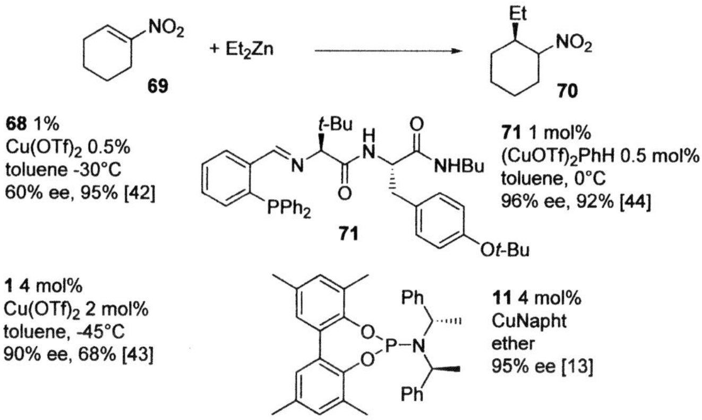

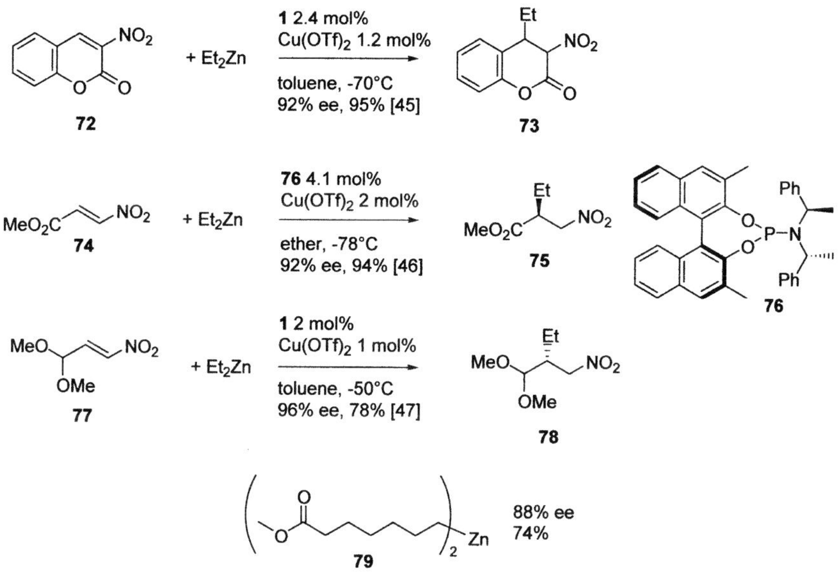

Scheme 8

functional group compatibility of the zinc reagent was illustrated with reagent **79**, which gave an adduct in high ee and with good chemical yield [47].

3
Rhodium-Catalyzed Arylation Reaction of Arylmetallic Reagents by Using External Chiral Ligands

The efficient conjugate addition-type arylation of activated olefins has not been achieved by the use of copper-catalyzed reactions. This situation was circum-

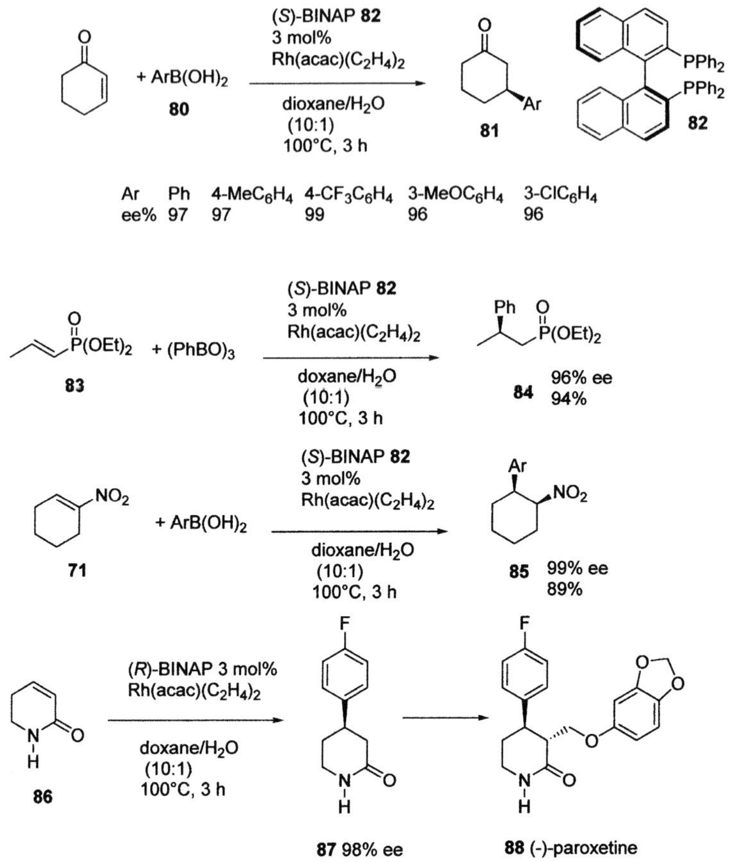

Scheme 9

vented by the use of the rhodium-catalyzed reaction, which was originally developed by Miyaura. The team of Hayashi and Miyaura succeeded in the arylation of cyclohexenone in a conjugate addition reaction of arylboronic acid. The arylation of cyclohexenone with arylboronic acid **80** under the catalysis of rhodium(I)–BINAP **82** proceeded very well to give 3-arylcyclohexanone **81** in excellent ee and yield (Scheme 9) [48]. The reaction is applicable to the arylation of vinylphosphonate **83**, nitroalkene **71**, and lactam **86** to give the corresponding arylated products in high enantioselectivity [49–51]. Application to the asymmetric synthesis of the pharmacologically potent compound paroxetine (**88**) illustrates the usefulness of the arylation reaction [51]. Mechanistic study revealed that phenylation of Rh(acac) to **90** is the essential step for the asymmetric arylation (Scheme 10) [52].

Arylmetallic reagents other than arylboronic acid were shown to be applicable in the rhodium–BINAP-catalyzed arylation. Aryltitanium reagent **93** con-

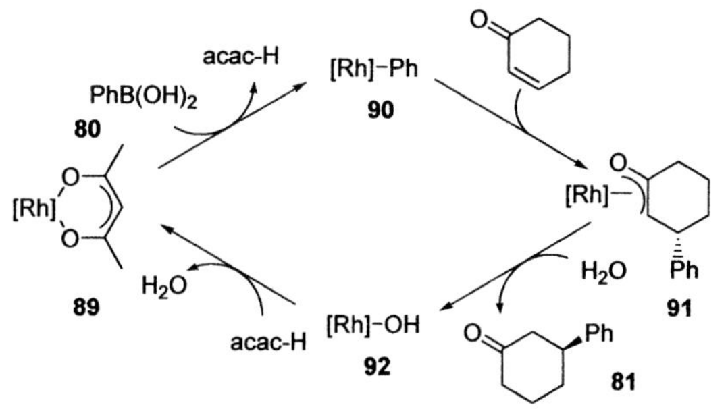

Scheme 10

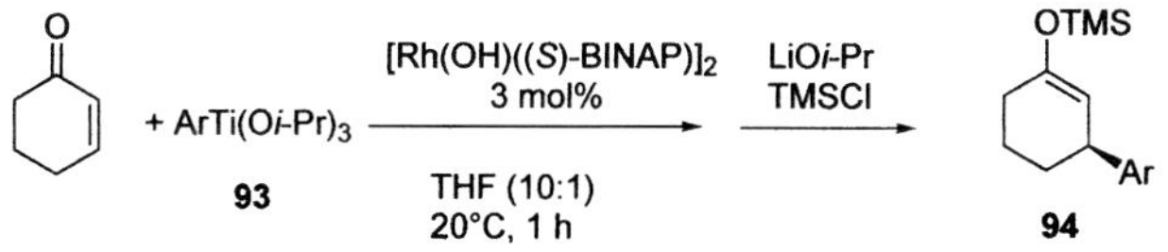

Ar	Ph	2-MeC$_6$H$_4$	4-FC$_6$H$_4$	4-MeOC$_6$H$_4$
ee%	99.5	94	99	99.8

Scheme 11

verted cyclohexenone to the silyl enol ether **94** with extremely high ee, which is useful for further standard synthetic manipulation under the anhydrous conditions (Scheme 11) [53]. Potassium aryltrifluoroborate was introduced by another group as an easily handled aryl group source that gave **81** in 98% ee [54]. Phosphines other than BINAP **82** were generally not applicable in the present rhodium-catalyzed arylation with the exceptions of chiral amidophosphine **95** (**81**: 95% ee) [55] and phosporamidite **96** (**81**: 89% ee) [56]. The phosphine **95** was claimed to work as a hemilabile ligand in the arylation [56]. Quite recently, arylsilane **97** was introduced as a stable and safe source of an aryl group which gave **81** in 98% ee [57]. Table 5 provides a useful overview and summary of the present status of the rhodium-catalyzed phenylation of cyclohexenone.

4
Radical Reaction by Using External Chiral Ligands

It is remarkable and impressive to find that stereochemistry in the conjugate addition of a radical species to an activated olefin is controlled by the chiral bisoxazoline ligand **98** to give a conjugate addition product **100** from **99** in quite high ee and diastereoselectivity (Scheme 12) [58]. Radical trapping by hydrogen abstraction was also shown to be possible in the reaction of **99** to **103**, which was controlled by the combination of a chiral alcohol **101** and achiral oxazolidinone **102** [59].

Table 5 Rhodium-catalyzed phenylation of cyclohexenone

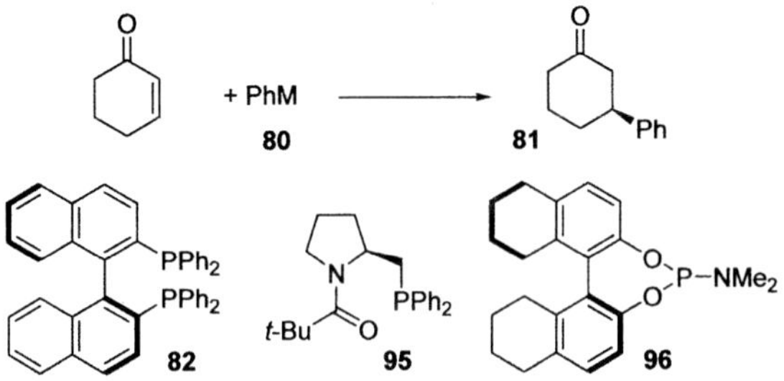

(*S*)-BINAP **82** 3 mol% Rh(acac)(C₂H₄)₂ PhB(OH)₃ 97% ee [48]	(*S*)-BINAP **82** 3 mol% Rh(acac)(C₂H₄)₂ PhTl(O-*i*-Pr)₃ **93** 99.5% ee [53]	(*S*)-BINAP **82** 3 mol% Rh(acac)(C₂H₄)₂ PhBF₃K 98% ee [54]
95 3 mol% Rh(acac)(C₂H₄)₂ PhB(OH)₃ 97% ee [55]	**96** 3 mol% Rh(acac)(C₂H₄)₂ PhB(OH)₃ 89% ee, quant [56]	(*S*)-BINAP **82** 3 mol% Rh(acac)(C₂H₄)₂ PhSi(OMe)₃ **97** 98% ee [57]

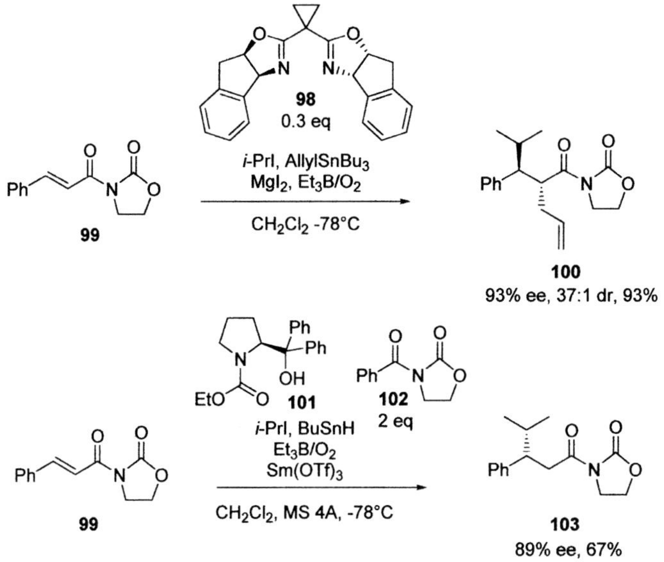

Scheme 12

5
Reaction of Organolithium Reagents by Using External Chiral Ligands

Asymmetric arylation of chiral oxazolines has provided a versatile method for the arylation of enoates. For example, **104** was arylated with an aryllithium re-

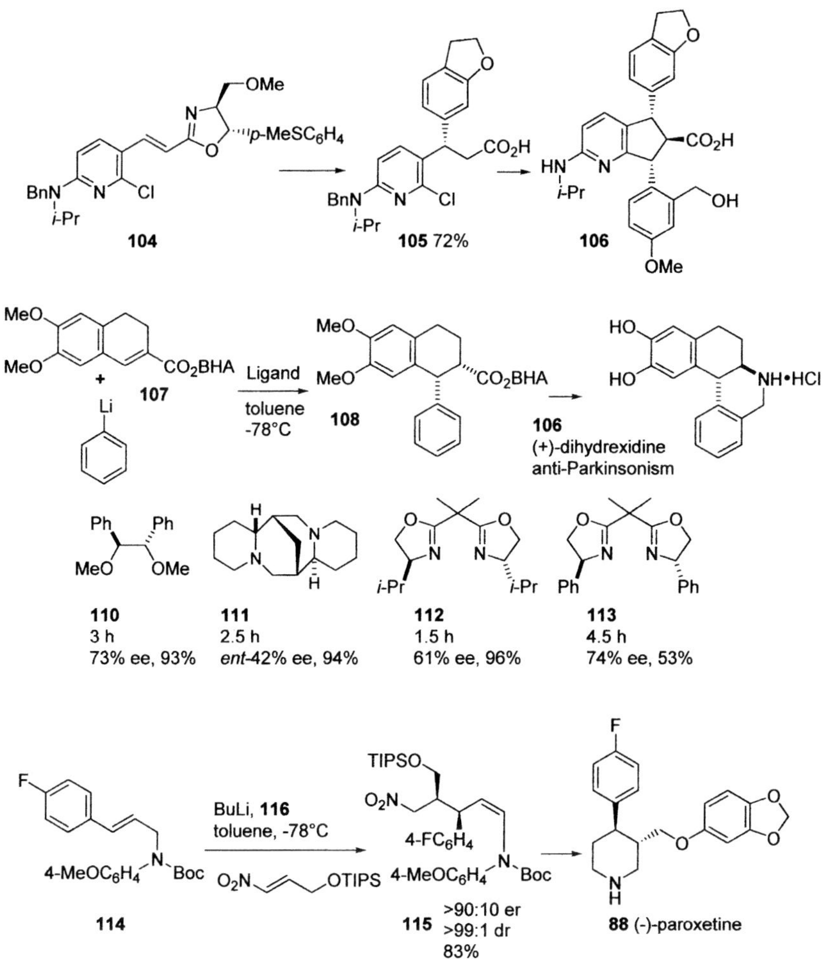

Scheme 13

agent to afford **105** that, after hydrolytic removal of the chiral auxiliary group, was converted to the selective endotherin A receptor antagonist **106** by a Merck–Banyu group (Scheme 13) [60]. A chiral ligand method was introduced by Tomioka's group, in which addition of phenyllithium to an enoate **107** was controlled by the chiral diether ligand **110** to give the adduct **108** in 73% ee [61]. The product **108** was converted to dopamine D1 agonist dihydrexidine (**109**). Sparteine (**111**) and chiral Boxes **112** and **113** were inferior to **110**. The rhodium-catalyzed arylation of the Hayashi method was unsuccessful for the phenylation of **107**.

Beak reported that sparteine (**111**) mediated the conjugate addition of lithiated allylamine **114** to give nitroalkene adduct **115** with high diastereoselectivity and enantioselectivity [62]. The product **115** was converted to paroxetine (**88**) by conventional synthetic manipulations.

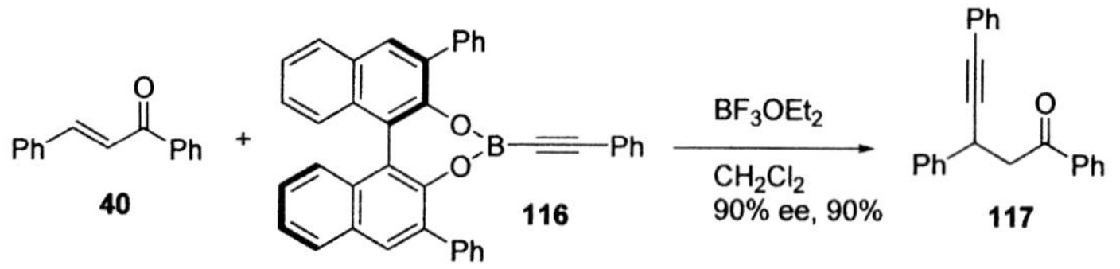

Scheme 14

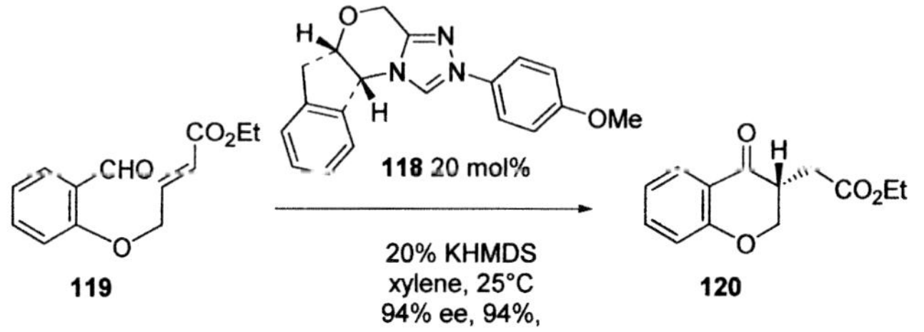

Scheme 15

6
Reaction of Chiral Organometallic Reagents

A binaphthol-based chirally modified acetylide **116** was shown to undergo conjugate addition to chalcone **40** to give adduct **117** in 90% ee (Scheme 14) [63]. Enones capable of achieving an s-*cis* conformation are claimed to be suitable substrates for this addition reaction.

7
Principal Alternatives

The use of species **118** for the umpolung of a carbonyl group, the Stetter reaction, was demonstrated for the intramolecular asymmetric conjugate addition of a formyl group to the enoate moiety of **119** to give the cyclization product **120** in high ee and yield (Scheme 15) [64].

References

1. Feringa BL (2000) Acc Chem Res 33:346; Tomioka K (2000) In: Otera J (ed) Modern carbonyl chemistry, Chap 12. Wiley-VCH, Germany; Sibi MP, Manyem S (2000) Tetrahedron 56:8033; Krause N, Hoffmann-Roder A (2001) Synthesis 171
2. Tomioka K (1990) Synthesis 541
3. Arnold LA, Imbos R, Mandoli A, de Vries AHM, Naasz R, Feringa BL (2000) Tetrahedron 56:2865
4. Hu X, Chen H, Zhang X (1999) Angew Chem Int 38:3518
5. Escher IH, Pfaltz A (2000) Tetrahedron 56:2879
6. Arena CG, Calabro G, Francio G, Faraone F (2000) Tetrahedron Asymmetry 11:2387
7. Choi YH, Choi JY, Yang HY, Kim YH (2002) Tetrahedron Asymmetry 13:801

8. Huttenloch O, Laxman E, Waldmann H (2002) Chem Commun 673; Huttenloch O, Laxman E, Waldmann H (2002) Chem Eur J 8:4767
9. Mandoli A, Arnold LA, Imbos R, de Vries AHM, Salvadori P, Feringa BL (2001) Tetrahedron Asymmetry 12:1929
10. Liang L, Au-Yeung TTL, Chan ASC (2002) Org Lett 4:3799
11. Reetz MT, Gosberg A, Moulin D (2002) Tetrahedron Lett 43:1189
12. Reetz MT, Maiwald P (2002) C R Chimie 5:341
13. Alexakis A, Benhaim C, Rosset S, Humam M (2002) J Am Chem Soc 124:5262
14. Dieguez M, Ruiz A, Claver C (2001) Tetrahedron Asymmetry 12:2895
15. Dieguez M, Pamies O, Net G, Ruiz A, Claver C (2002) J Mol Cat A Chem 185:11
16. Pamies O, Net G, Ruiz A, Claver C, Woodward S (2000) Tetrahedron Asymmetry 11:3181
17. Alexakis A, Burton J, Vastra J, Benhaim C, Fournioux X, Van den Heuvel A, Leveque JM, Maze F, Rosset S (2000) Eur J Org Chem 4011
18. Delapierre G, Constantieux T, Brunel JM, Buono G (2000) Eur J Org Chem 2507
19. Taira S, Crepy KVL, Imamoto T (2002) Chirality 14:386
20. Kinahan TC, Tye H (2001) Tetrahedron Asymmetry 12:1255
21. Morimoto T, Yamaguchi Y, Suzuki M, Saitoh A (2000) Tetrahedron Lett 41:10025
22. Nakagawa Y, Matsumoto K, Tomioka K (2000) Tetrahedron 56:2857; Tomioka K, Nakagawa Y (2000) Heterocycles 52:95
23. Degrado SJ, Mizutani H, Hoveyda AH (2001) J Am Chem Soc 123:755
24. Guillen F, Winn CL, Alexakis A (2001) Tetrahedron Asymmetry 12:2083
25. Pytkowicz J, Sylvain R, Mangeney P (2001) Tetrahedron Asymmetry 12:2087
26. Knopff O, Alexakis A (2002) Org Lett 4:3835
27. Alexakis A, Trevitt GP, Bernardinelli G (2001) J Am Chem Soc 123:4358
28. Imbos R, Brilman MHG, Pineschi M, Feringa BL (1999) Org Lett 1:623; Imbos R, Minnaard AJ, Feringa BL (2001) Tetrahedron 57:2485
29. Braga AL, Silva SJN, Ludtke DS, Drekener RL, Silveria CC, Rocha JBT, Wessjohann LA (2002) Tetrahedron Lett 43:7329
30. Yang M, Zhou ZY, Chan ASC (2000) Chem Commun 115
31. Shintani R, Fu GC (2002) Org Lett 4:3699
32. Yin Y, Li X, Lee DS, Yang TK (2000) Tetrahedron Asymmetry 11:3329
33. Tong PE, Li P, Chan ASC (2001) Tetrahedron Asymmetry 12:2301
34. Shadakshari U, Nayak SK (2001) Tetrahedron 57:8185
35. Wakimoto I, Tomioka Y, Kawanami Y (2002) Tetrahedron 58:8095
36. Martorell A, Naasz R, Feringa BL, Pringle PG (2001) Tetrahedron Asymmetry 12:2497
37. Bennett SMW, Brown SM, Cunningham A, Dennis MR, Muxworthy JP, Oakley MA, Woodward S (2000) Tetrahedron 56:2847; Borner C, Konig WA, Woodward S (2001) Tetrahedron Lett 42:327; Garcia-Ruiz V, Woodward S (2002) Tetrahedron Asymmetry 13:2177
38. Mizutani H, Degrado SJ, Hoveyda AH (2002) J Am Chem Soc 124:779
39. Alexakis A, Benhaim C (2001) Tetrahedron Asymmetry 12:1151
40. Soeda T, Fujihara H, Tomioka K (unpublished result); Fujihara H, Nagai K, Tomioka K (2000) J Am Chem Soc 122:12055; Nagai K, Fujihara H, Kuriyama K, Yamada K, Tomioka K (2002) Chem Lett 8
41. Ongeri S, Piarulli U, Jackson RFW, Gennari C (2001) Eur J Org Chem 803
42. Alexakis A, Benhaim C (2000) Org Lett 2:2579
43. Duursma A, Minnaard AJ, Feringa BL (2002) Tetrahedron 58:5773
44. Luchaco-Cullis CA, Hoveyda AH (2002) J Am Chem Soc 124:8192
45. Versleijen JPG, Van Leusen AM, Feringa BL (1999) Tetrahedron Lett 40:5803
46. Rimkus A, Sewald N (2003) Org Lett 5:79
47. Duursma A, Minnaard AJ, Feringa BL (2003) J Am Chem Soc 125:3700
48. Takaya Y, Ogasawara M, Hayashi T, Sakai M, Miyaura N (1998) J Am Chem Soc 120:5579
49. Hayashi T, Senda T, Takaya Y, Ogasawara M (1999) J Am Chem Soc 121:11591
50. Hayashi T, Senda T, Ogasawara M (2000) J Am Chem Soc 122:10716
51. Senda T, Ogasawara M, Hayashi T (2001) J Org Chem 66:6852

52. Hayashi T, Takahashi M, Takaya Y, Ogasawara M (2002) J Am Chem Soc 124:5052
53. Hayashi T, Tokunaga N, Yoshida K, Han JW (2002) J Am Chem Soc 124:12102
54. Pucheault M, Darses S, Genet JP (2002) Tetrahedron Lett 43:6155
55. Kuriyama M, Nagai K, Yamada K, Miwa Y, Taga T, Tomioka K (2002) J Am Chem Soc 124:8932; Kuriyama M, Tomioka K (2001) Tetrahedron Lett 42:921
56. Boiteau JG, Imbos R, Minnaard AJ, Feringa BL (2003) Org Lett 5:681
57. Oi S, Taira A, Honma Y, Inoue Y (2003) Org Lett 5:97
58. Sibi MP, Chen J (2001) J Am Chem Soc 123:9472
59. Sibi MP, Manyem S (2002) Org Lett 4:2929
60. Kato Y, Niiyama K, Nemoto T, Jona H, Akao A, Okada S, Song ZJ, Zhao M, Tsuchiya Y, Tomimoto K, Mase T (2002) Tetrahedron 58:3409
61. Asano Y, Yamashita M, Nagai K, Kuriyama M, Yamada K, Tomioka K (2001) Tetrahedron Lett 42:8493
62. Johnson TA, Curtis MD, Beak P (2001) J Am Chem Soc 123:1004
63. Chong JM, Shen L, Taylor NJ (2000) J Am Chem Soc 122:1822
64. Kerr MS, Read de Alaniz J, Rovis T (2002) J Am Chem Soc 124:10298

Supplement to Chapter 34.2
Protonation of Enolates

Akira Yanagisawa

Department of Chemistry, Faculty of Science, Chiba University,
Inage, Chiba 263–8522, Japan
e-mail: ayanagi@scichem.s.chiba-u.ac.jp

Keywords: Protonation, Metal enolates, Chiral proton sources, Achiral proton sources

1   Introduction 125

2   Mechanism of Catalysis 127

3   Catalytic Protonation of Metal Enolates 127

4   Catalytic Protonation of Ammonium Enolates 130

5   Principal Alternative 130

References 131

1
Introduction

The chemistry of asymmetric protonation of enols or enolates has further developed since the original review in *Comprehensive Asymmetric Catalysis* [1]. Numbers of literature reports of new chiral proton sources have emerged and several reviews [2–6] cover the topics up to early 2001. This chapter concentrates on new examples of catalytic enantioselective protonation of prochiral metal enolates (Scheme 1). Compounds 1–41 [7–45] shown in Fig. 1 are the chiral proton sources or chiral catalysts reported since 1998 which have been employed for the asymmetric protonation of metal enolates. Some of these have been successfully utilized in the catalytic version.

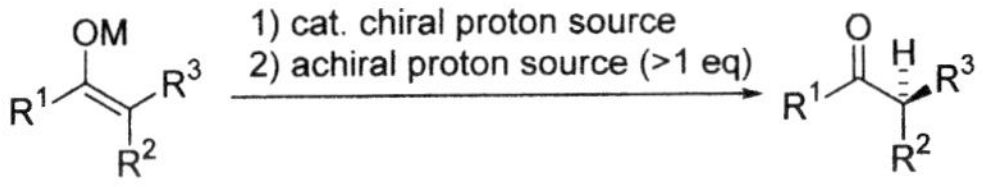

Scheme 1

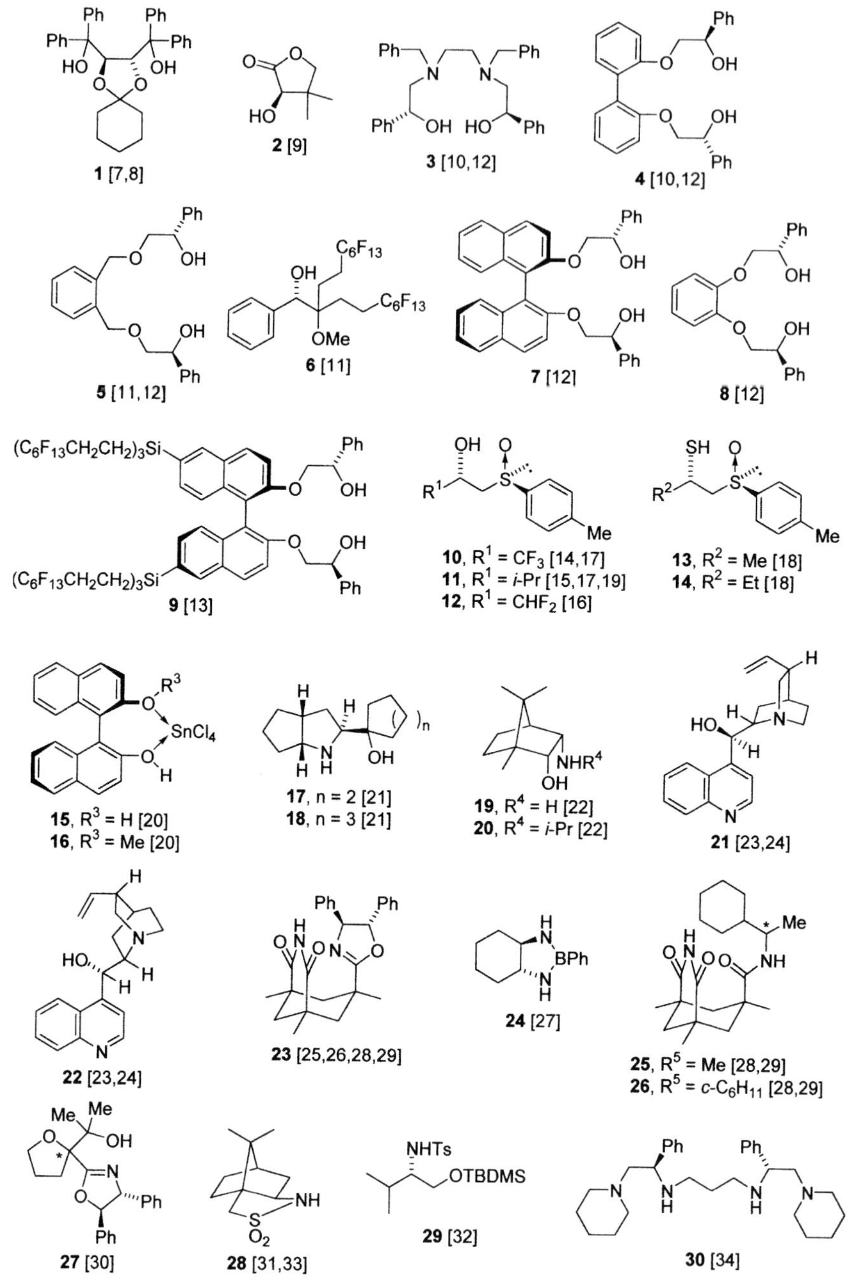
1 [7,8]
2 [9]
3 [10,12]
4 [10,12]
5 [11,12]
6 [11]
7 [12]
8 [12]
9 [13]
10, R1 = CF3 [14,17]
11, R1 = i-Pr [15,17,19]
12, R1 = CHF2 [16]
13, R2 = Me [18]
14, R2 = Et [18]
15, R3 = H [20]
16, R3 = Me [20]
17, n = 2 [21]
18, n = 3 [21]
19, R4 = H [22]
20, R4 = i-Pr [22]
21 [23,24]
22 [23,24]
23 [25,26,28,29]
24 [27]
25, R5 = Me [28,29]
26, R5 = c-C6H11 [28,29]
27 [30]
28 [31,33]
29 [32]
30 [34]

Fig. 1

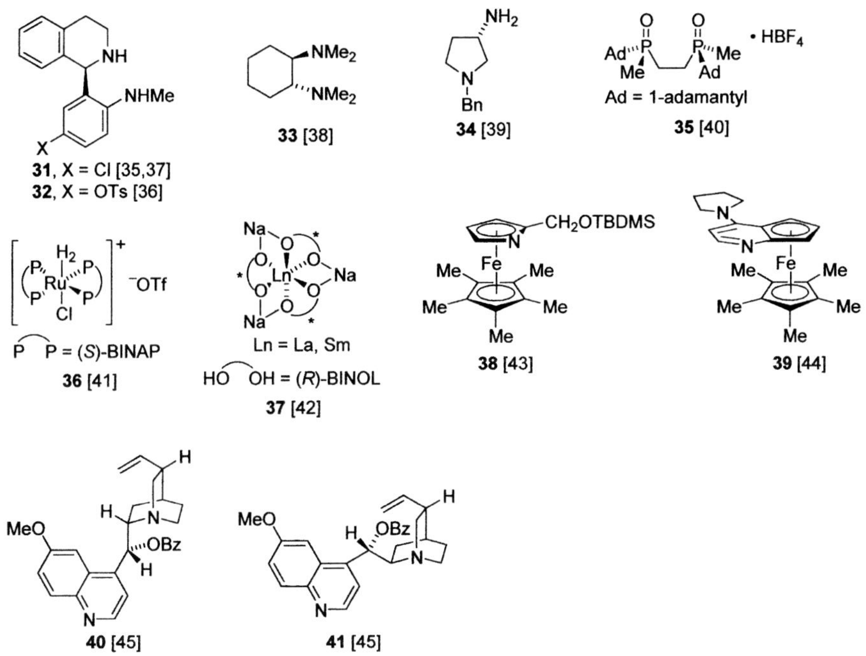

31, X = Cl [35,37]
32, X = OTs [36]

33 [38]

34 [39]

35 [40]

P P = (S)-BINAP

36 [41]

Ln = La, Sm

HO OH = (R)-BINOL

37 [42]

38 [43]

39 [44]

40 [45]

41 [45]

Fig. 1 (continued)

2
Mechanism of Catalysis

Several new methods for the asymmetric protonation of metal enolates have appeared; however, the catalytic mechanisms are fundamentally the same as that described in Scheme 2 of the 1st edition.

3
Catalytic Protonation of Metal Enolates

Several new catalytic asymmetric protonations of metal enolates under basic conditions have been published to date. In those processes, reactive metal enolates such as lithium enolates are usually protonated by a catalytic amount of chiral proton source and a stoichiometric amount of achiral proton source. Vedejs et al. reported a catalytic enantioselective protonation of amide enolates [35]. For example, when lithium enolate **43**, generated from racemic amide **42** and s-BuLi, was treated with 0.1 equivalents of chiral aniline **31** followed by slow addition of 2 equivalents of *tert*-butyl phenylacetate, (R)-enriched amide **42** was obtained with 94% ee (Scheme 2). In this reaction, various achiral acids were

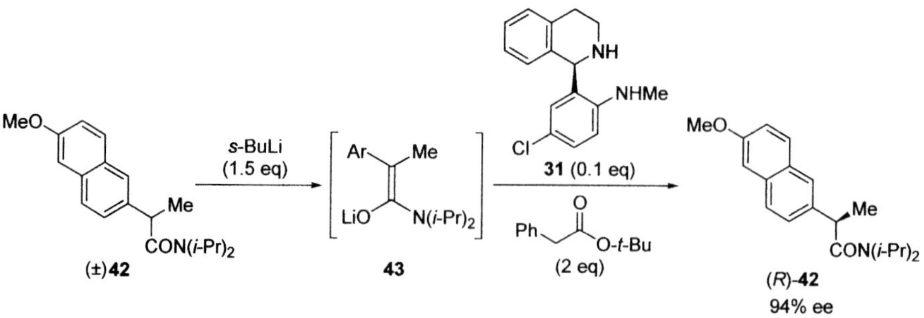

Scheme 2

investigated and it was found that the optimal pK_a value for the achiral proton source should be near that of the chiral proton source; however, significant enantioselectivities were observed over a broad pK_a (DMSO) range of 22–30.

Our research group developed catalytic enantioselective protonations of preformed enolates of simple ketones with (S,S)-imide 23 or chiral imides 25 and 26 based on a similar concept [29]. For catalytic protonation of a lithium enolate of 2-methylcyclohexanone, chiral imide 26, which possesses a chiral amide moiety, was superior to (S,S)-imide 23 as a chiral acid and the enolate was protonated with up to 82% ee.

Chiral α-sulfinyl alcohol (S,R$_S$)-11 was also shown to be a promising chiral proton donor in catalytic protonation of 2-methyl tetralone enolate by Asensio's group [19].

In contrast, Koga and coworkers found that enantioselective protonation of lithium enolates of 2-substituted-1-tetralones occurred with a catalytic amount of chiral tetraamine 30 in the presence of water as an achiral proton source [34]. This protonation system is noteworthy, since high enantioselectivities are observed notwithstanding the existence of a large excess of water.

Takeuchi and coworkers have achieved the catalytic enantioselective protonation of a samarium enolate 45 in the THF/FC-72 [$F_3C(CF_2)_4CF_3$] biphasic system using a C_2-symmetric chiral diol 5 (DHPEX) or a fluorinated chiral alcohol 6 as a catalyst and a fluorinated achiral alcohol 46 (Scheme 3) [11]. The fluorinated biphasic system was more effective than THF alone, and enantioselectivities near maximum values were obtained in the reaction. In addition, it was unnecessary to add the achiral alcohol 46 slowly to the reaction mixture.

Heterobimetallic asymmetric complexes developed by Shibasaki et al. are known as effective catalysts for asymmetric Michael additions. They achieved a catalytic asymmetric protonation in Michael additions of thiols to α,β-unsaturated carbonyl compounds using LaNa$_3$tris(binaphthoxide) and SmNa$_3$tris(binaphthoxide) complexes (SmSB) 37 [42]. For instance, treatment of thioester 48 with 4-tert-butyl(thiophenol) and 0.1 equivalents of (R)-SmSB 37 (Ln=Sm) in CH$_2$Cl$_2$ at -78°C gave the adduct 49 in 93% ee and in 86% yield (Scheme 4). The high enantiomeric ratio is considered to be attributable to an

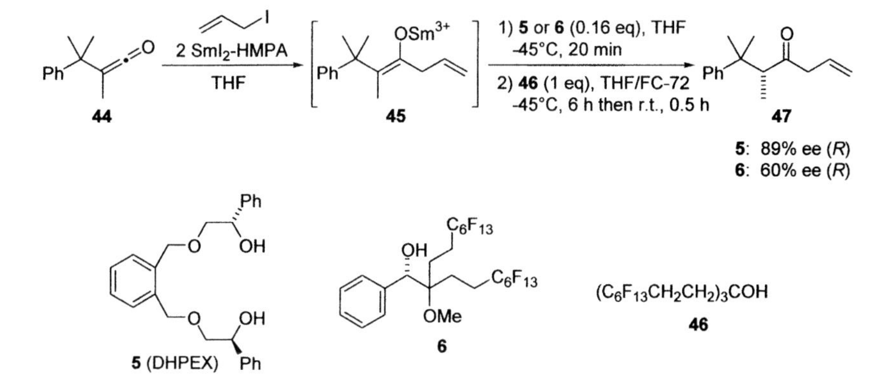

5: 89% ee (R)
6: 60% ee (R)

5 (DHPEX)   Ph

6

$(C_6F_{13}CH_2CH_2)_3COH$

46

Scheme 3

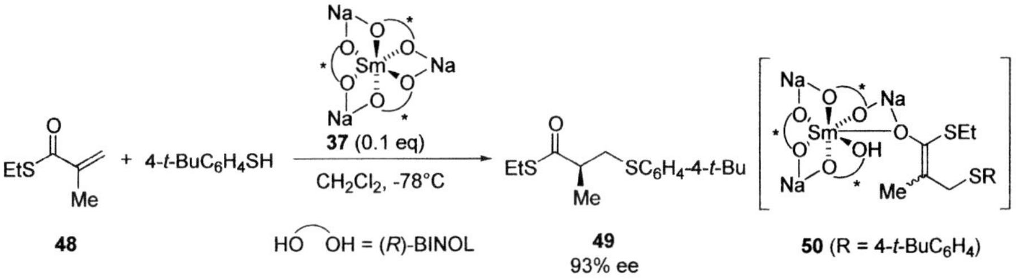

Scheme 4

acidic OH portion of the intermediate enolate **50** generated by the Michael addition.

Muzart and coworkers have reported a new catalytic enantioselective protonation of prochiral enolic species generated by palladium-induced cleavage of β-ketoesters or enol carbonates of α-alkylated 1-indanones and 1-tetralones [21]. Among the various (S)-β-aminocycloalkanols examined, **17** and **18** were effective chiral catalysts for the asymmetric reaction and (R)-enriched α-alkylated 1-indanones and 1-tetralones were obtained with up to 72% ee. In some cases, the reaction temperature affected the ee.

Later, the same group showed that a racemic open chain benzyl β-ketoester was also converted to the corresponding optically active ketone according to a similar procedure using cinchona alkaloids **21** or **22** as a chiral proton source [23].

Yamamoto et al. reported full research details on catalytic enantioselective protonation under acidic conditions in which prochiral trialkylsilyl enol ethers and ketene bis(trialkyl)silyl acetals were protonated by a catalytic amount of Lewis acid assisted Brønsted acid (LBA **15** or **16**) and a stoichiometric amount of 2,6-dimethylphenol as an achiral proton source [20].

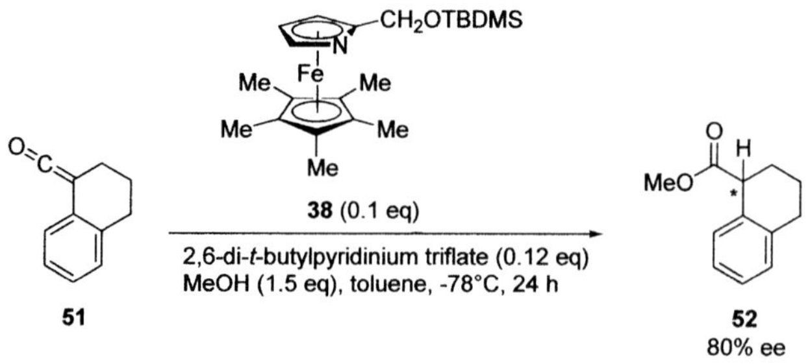

Scheme 5

4
Catalytic Protonation of Ammonium Enolates

Addition of alcohols to ketenes in the presence of chiral tertiary amine compounds is a convenient way of preparing optically active esters via a protonation of chiral ammonium enolates generated in situ. Fu and coworkers achieved a catalytic version of this process by employing planar-chiral azaferrocene **38** as chiral catalyst [43]. For example, treatment of ketene **51** with a mixture of the catalyst **38** (0.1 equiv), 2,6-di-*tert*-butylpyridinium triflate (0.12 equiv), and MeOH (1.5 equiv) in toluene at –78°C afforded the methyl ester **52** with 80% ee (Scheme 5). The same group further developed an asymmetric addition of pyrroles to ketenes catalyzed by planar-chiral 4-(pyrrolidino)pyridine **39** leading to enantioenriched *N*-acylpyrroles with up to 98% ee [44]. A transformation of silylketenes into chiral nonracemic α-silylthioesters was also accomplished by Simpkins et al. according to a similar methodology using cinchona alkaloid catalysts **40** and **41** [45].

5
Principal Alternative

Several processes promoted by a stoichiometric amount of chiral acids or chiral ligands have been developed for enantioselective protonation of metal enolates: for example, protonation of preformed lithium enolates generated by deprotonation of the corresponding ketones and esters or by reaction of the corresponding silyl enol ethers or enol acetates with RLi [9, 14–18, 25–28, 30, 36–39, 41]; protonation of boron enolates generated from glycine derivatives with cinchona alkaloids **21** and **22** [24]; a SmI_2-mediated reduction of α-heterosubstituted α-alkyl or α-arylketones or lactone with chiral diols **3**, **4**, and **7–9** [10, 12, 13], a SmI_2-induced reductive coupling of chiral 2-alkyl acrylates or achiral methyl methacrylate with ketones in the presence of chiral proton sources **28** and **29** [31–33]; and deracemization of 2-alkylcyclohexanones with chiral diol **1** in basic suspension media [7, 8]. Silyl enol ethers can be enantioselectively protonated by P-chirogenic diphosphine oxide-Brønsted acid complex **35** under acidic conditions [40]. Effects of reaction temperature have been studied on asymmetric photodeconjugation of chiral ammonium α,β-unsaturated carboxylates leading to the corresponding optically active β,γ-unsaturated esters [22]. Racemic α-substituted carboxylic acids can be deracemized in the presence of biocatalysts [46, 47]. Asymmetric hydrolysis of enol esters with some enzymes [48] or plants [49] is also a beneficial way to optically active 2-alkylcyclohexanones.

References

1. Yanagisawa A, Yamamoto H (1999) In: Jacobsen EN, Pfaltz A, Yamamoto H (eds) Comprehensive asymmetric catalysis III, Chap 34.2. Springer, Berlin Heidelberg New York, p 1295
2. Ebbers EJ, Ariaans GJA, Houbiers JPM, Bruggink A, Zwanenburg B (1997) Tetrahedron 53:9417
3. Krause N, Ebert S, Haubrich A (1997) Liebigs Ann. Recueil 2409
4. Calmes M, Daunis J (1999) Amino Acids 16:215
5. Eames J, Weerasooriya N (2001) Tetrahedron Asymmetry 12:1
6. Dalko PI, Moisan L (2001) Angew Chem Int Ed 40:3727
7. Kaku H, Ozako S, Kawamura S, Takatsu S, Ishii M, Tsunoda T (2001) Heterocycles 55:847
8. Kaku H, Takaoka S, Tsunoda T (2002) Tetrahedron 58:3401
9. Calmès M, Glot C, Martinez J (2001) Tetrahedron Asymmetry 12:49
10. Mikami K, Yamaoka M, Yoshida A, Nakamura Y, Takeuchi S, Ohgo Y (1998) Synlett 607
11. Takeuchi S, Nakamura Y, Ohgo Y, Curran DP (1998) Tetrahedron Lett 39:8691
12. Nakamura Y, Takeuchi S, Ohgo Y, Yamaoka M, Yoshida A, Mikami K (1999) Tetrahedron 55:4595
13. Nakamura Y, Takeuchi S, Ohgo Y, Curran DP (2000) Tetrahedron 56:351
14. Asensio G, Aleman PA, Domingo LR, Medio-Simón M (1998) Tetrahedron Lett 39:3277
15. Asensio G, Alemán P, Cuenca A, Gil J, Medio-Simón M (1998) Tetrahedron Asymmetry 9:4073
16. Asensio G, Aleman P, Gil J, Domingo LR, Medio-Simon M (1998) J Org Chem 63:9342
17. Asensio G, Cuenca A, Gaviña P, Medio-Simón M (1999) Tetrahedron Lett 40:3939
18. Asensio G, Gaviña P, Cuenca A, de Arellano MCR, Domingo LR, Medio-Simón M (2000) Tetrahedron Asymmetry 11:3481
19. Asensio G, Gil J, Alemán P, Medio-Simon M (2001) Tetrahedron Asymmetry 12:1359
20. Nakamura S, Kaneeda M, Ishihara K, Yamamoto H (2000) J Am Chem Soc 122:8120
21. Aboulhoda SJ, Reiners I, Wilken J, Hénin F, Martens J, Muzart J (1998) Tetrahedron Asymmetry 9:1847
22. Hénin F, Létinois S, Muzart J (2000) Tetrahedron Asymmetry 11:2037
23. Roy O, Diekmann M, Riahi A, Hénin F, Muzart J (2001) Chem Commun 533
24. O'Donnell MJ, Drew MD, Cooper JT, Delgado F, Zhou C (2002) J Am Chem Soc 124:9348
25. Yanagisawa A, Kikuchi T, Yamamoto H (1998) Synlett 174
26. Yanagisawa A, Kikuchi T, Kuribayashi T, Yamamoto H (1998) Tetrahedron 54:10253
27. Yanagisawa A, Inanami H, Yamamoto H (1998) Chem Commun 1573
28. Yanagisawa A, Kikuchi T, Watanabe T, Yamamoto H (1999) Bull Chem Soc Jpn 72:2337
29. Yanagisawa A, Watanabe T, Kikuchi T, Yamamoto H (2000) J Org Chem 65:2979
30. Yanagisawa A, Matsuzaki Y, Yamamoto H (2001) Synlett 1855
31. Xu M-H, Wang W, Lin G-Q (2000) Org Lett 2:2229
32. Wang W, Xu M-H, Lei X-S, Lin G-Q (2000) Org Lett 2:3773
33. Xu M-H, Wang W, Xia L-J, Lin G-Q (2001) J Org Chem 66:3953
34. Yamashita Y, Emura Y, Odashima K, Koga K (2000) Tetrahedron Lett 41:209

35. Vedejs E, Kruger AW (1998) J Org Chem 63:2792
36. Vedejs E, Kruger AW, Suna E (1999) J Org Chem 64:7863
37. Vedejs E, Kruger AW, Lee N, Sakata ST, Stec M, Suna E (2000) J Am Chem Soc 122:4602
38. Eames J, Weerasooriya N (2000) Tetrahedron Lett 41:521
39. Flinois K, Yuan Y, Bastide C, Harrison-Marchand A, Maddaluno J (2002) Tetrahedron 58:4707
40. Matsukawa S, Imamoto T (2000) J Am Chem Soc 122:12659
41. Nishibayashi Y, Takei I, Hidai M (1999) Angew Chem Int Ed 38:3047
42. Emori E, Arai T, Sasai H, Shibasaki M (1998) J Am Chem Soc 120:4043
43. Hodous BL, Ruble JC, Fu GC (1999) J Am Chem Soc 121:2637
44. Hodous BL, Fu GC (2002) J Am Chem Soc 124:10006
45. Blake AJ, Friend CL, Outram RJ, Simpkins NS, Whitehead AJ (2001) Tetrahedron Lett 42:2877
46. Chadha A, Baskar B (2002) Tetrahedron Asymmetry 13:1461
47. Kato D, Mitsuda S, Ohta H (2002) Org Lett 4:371
48. Hirata T, Shimoda K, Kawano T (2000) Tetrahedron Asymmetry 11:1063
49. Bruni R, Fantin G, Medici A, Pedrini P, Sacchetti G (2002) Tetrahedron Lett 43:3377

Subject Index

Acetyl pyridines   9
Acyloxyborane, chiral   100
AD/*N*-oxidation   33
Adrenergic receptor agonists   59
Aflatoxin B lactone   84, 85
Alcohol-nickel catalysts, bidentate   114
Alcohols, homoallylic   50
–, nonracemic tertiary homoallylic   105
–, secondary   7, 8
Aldehydes, allyltributyltin, enantioselective
    addition   101
–, enantioselective allylation   100
Alkene, silica-anchored tetrasubstituted   31
Alkenes, AD, ligands   23
–, β-aminoalcohols   44
–, Baylis-Hillman   25
–, DHQD ligands   24
–, enantioselective hydrogenation   4
–, substituted, AA   49
–, unfunctionalized   1
–, zirconocenes   2
cis-Alkenes   22
2-Alkyl-1,3-diketones   16
Alkylation   109
–, allylic   73
Allenyltributylstannane, aldehydes   104
Allylamine, lithiated   121
Allylation   73, 97
–, Lewis base catalysts   103
Allylic alcohols, γ-fluoroalkylated   83
Allylic substitution   73
Allylsilanes   97
Allylstannates   97, 104
Allyltributyltin, enantioselective addition to
    aldehydes   101
Allyltrichlorosilanes   103, 104
Amastatin   64
Amide enolates   127
Amino acids, derivatives   83
– –, protected   79
Amino alcohols   13
–, chiral   56
Aminocyclohexitols   66
2-Amino-1,2-diphenylethanol   66

Aminohydroxylation, asymmetric (AA)   43
–, –, diastereoselectivity   67
–, –, procedures   45
Ammonium enolates, catalytic
    protonation   130
Anthraquinone   22
Aryl esters, reversal of regioselectivity   47
Arylacrylates, heteroaromatic   56, 58
Arylation, rhodium-BINAP-catalyzed   118
Arylglycines   61
Aryllithium   120
Arylmetallic reagents, rhodium-catalyzed
    arylation   117
Atropisomers, non-biaryl   77
Azetidinone   63

Barbituric acid   82
Baylis-Hillman alkenes   25
– –, aminohydroxylation, diastereo-
    selective   68
Benzylamine   76
Benzylic alcohol   48
Benzylic amide   48
Bestatin   64
Bisalkaloid ligands   38
Bisphosphine, copper-catalyzed organozinc
    addition   113
Bis(oxazolinyl)phenyl rhodium(III)   101
Bis(*N*-tosylamino)phosphines   85
Borane reagents   7, 8
Borohydride   7
Brintzinger-type zirconocene cpmlexes   1
N-Bromoacetamide   46
Brønsted acids, chiral   105
BSA   79, 80, 88

Callipeltoside A   86
Carbamate-AA   51
Carbanucleosides   89
Carbene, chiral heterocyclic   1
Carbonyl compounds, allylic tin
    reagents   105
– –, asymmetric allylation   97, 100
– –, Lewis base catalyzed   103

Catecholborane   12
CBS reduction   7, 8
Chalcone, diethylzinc   115
Chemoselectivity, AA process   54
Chloramine-T   46
N-Chloro-N-metallosulfonamides   44
Cinchona alkaloids   21–26, 43, 47, 51
– –, AD, silica gel-supported   37
Cinnamates   47, 55, 56
Cobalt complexes, β-keto iminato   14
Conjugate addition   110
Cooxidants   27, 44
Copper   109
Copper triflate   110
Crabtree-type complexes   1
Crotyltrichlorosilanes   103
Cyclic substrates, Pd-catalyzed   81
Cycloalkenyl series   80
Cyclohexadienone   113
2,5-Cyclohexadienylsilanes, AA   68
Cyclohexenone   111, 112
Cyclomarin A   60
Cyclopentadienylruthenium   80
Cyclopentobarbital   82
Cyclophellitol   82

Dehydrotubifoline   82
Diethylzinc   111
Dihydroquinine (DHQ)   30
Dihydroxylation, asymmetric (AD)   21, 44
– –, molecular oxygen   26
– –, pH   27
–, catalytic cycles, homogeneous
   conditions   24
1,3-Diketones, stereoselective reduction   14
Dimethyl malonate/BSA   79, 80, 88
Dimethylzinc   110
Diphenylphthalazine   22
1,3-Diphenylprop-2-enyl acetate   77
1,3-Diphenylpropenyl ester   75, 76
1,3-Diphenylprop-2-enyl
   ethylcarbonate   80
Diphenylpyrazinopyridazine   22

Endotherin A   121
Enoates, arylation   120
–, prochiral   125
–, protonation   125, 127
Enones, acyclic, Ni/Cu-catalyzed   114
–, cyclic   110
Esters, conjugated   62
–, α,β,-unsaturated   50
Ethambutol   86

1-Ethyl-3-methylimidazolium
   tetrafluoroborate   34
3-Ethylcyclohexanone   110
Exochelin   60

Fenchone   77
Flavin hydrogenperoxide   27

Galanthamine   82

Heck/AD   33
Hydroboration   7
–, metal complexes, catalysis   12
Hydrogenation, zirconocene-catalyzed   2
4-Hydroxy-2-butenoates   50

Imidazolylidine, oxazoline   3
Imides, enolates   128
1-Indanones   129
Iridium   73, 87
– complexes, chiral   1, 3
Isopropyl cinnamate   55

Ketene silyl acetal   78
Ketenes, pyrroles   130
β,-Ketoesters   129
Ketones, aromatic   12
–, α-hetero substituted aliphatic   8

Lactacystin   65
Lewis acid/base catalysts, allylation   100,
   103
Ligands, chiral, immobilization   34
–, N,S-   78
–, PHAL   50, 62
–, P,N-   75
Lithium enolates   127
Loracarbef   64, 66

Malyngolide   86
Medermycin   65
MeO-PEG   38
Metal enolates, protonation   125
α-Methoxyacetophenone   14
N-Methylmorpholine N-oxide (NMO)   27
2-Methyl-3-phenyl-2-butene,
   dihydroxylation   26
Methylstilbene   5
α-Methyl styrene, asymmetric
   dihydroxylation   31
Molybdenum   73, 86
MOP   79

NaBH₄   14
2-Naphthoate   100
Naphthylacrylate   57
Nickel boride   10
Nitroalkenes   109
–, copper-catalyzed   116
Nitrogen source, AA reaction   45

Olefins, activated, conjugate addition   119
–, electrophilically activated   109
Organolithium   109
– reagents, external chiral ligands   120
Organometallic reagents, chiral   122
Organozinc   109
– reagents, copper-catalyzed conjugate
   addition   110
Osmaazetidine   53
Osmium, immobilization   30
Osmium azaglycolate   53
Osmium tetroxide   21, 43
Osmium(VI) bis(glycolate)   25
Oxazaborolidines, hydroboration
   catalyzed   7
–, immobilization   10
Oxazolines, chiral, arylation   120

Palladium   73
Palladium-catalyzed substitution,
   mechanism   74
PHAL ligands   50, 62
Phenylisoserine   25
Phenyllithium   121
Phosphine   109
– /oxazoline complexes, iridium   3
Phosphinite, oxazoline   3
Phosphinopyrrolyl, oxazoline   3
Phosphonates, unsaturated   57
Phosphoramidite   110
–, Taddol-based   115, 116
Phosphorus   109
Phthalazine   22
Pinacol   102
Polyketide synthesis   100

Potassium ferricyanide cooxidant   27
Propargyltrichlorosilane   103
Propiophenone   13
Protonation   125
Pyrimidine   22
Pyrrolidines   88

Quinuclidine   35

Radical addition, external chiral
   ligands   119
Rhodium   118
– BINAP-catalyzed arylation   118
–, cationic chiral   102
Ruthenium   73

Samarium enolate   128
Selenoxides   29
Serine, dihydroxyphenyl, fluorinated   59
Silver(I) catalyst, chiral   102
trans-Stilbene, AD   35, 36
Styrenes   5, 47, 61

Tetraallyltin   105
Tetralone enolate   128
Tetralones   90, 128, 129
Tetraponerines   89
Titanium complexes, chiral   100
Titanocene complex   2
Trioxoimidoosmium(VIII)   52
Tubifoline   82

Vanadyl acetylacetonate   28
Vigabatrin   86
cis-Vinyl epoxides   103
2-Vinylfuran   61
Vinylglycidols   86
Vitamin E   86

Zankiren   64
Zinc enolate   79
Zirconocene complexes, Brintzinger-type   1
– –, chiral cationic   2

Printing: Strauss GmbH, Mörlenbach
Binding: Schäffer, Grünstadt